Sustainable Development and the Limitation of Growth

Future Prospects for World Civilization

Victor I. Danilov-Danil'yan, Kim S. Losev, and
Igor E. Reyf

Sustainable Development and the Limitation of Growth

Future Prospects for World Civilization

 Springer

Published in association with
Praxis Publishing
Chichester, UK

Dr Victor I. Danilov-Danil'yan
Corresponding Member of the Russian Academy of Sciences; Director of the Water
    Institute at the Russian Academy of Sciences
Moscow
Russia

Professor Dr Kim S. Losev
Ecologist/Hydrologist
All Russian Institute of Scientific & Technical Information of the Russian Academy of
    Sciences
Moscow
Russia

Igor E. Reyf
Journalist
Frankfurt am Main
Germany

Translator: Associate Professor Vladimir Tumanov, Department of Modern Languages
and Literatures, University of Western Ontario, London, Ontario, Canada

---

SPRINGER–PRAXIS BOOKS IN ENVIRONMENTAL SCIENCES
SUBJECT *ADVISORY EDITOR*: John Mason B.Sc., M.Sc., Ph.D.

ISBN 978-3-540-75249-3 Springer Berlin Heidelberg New York

Springer is part of Springer-Science + Business Media (springer.com)

Library of Congress Control Number: 2008939852

Cover design: Jim Wilkie
Translation editor: Dr Donald Rapp
Project management: OPS Ltd, Gt Yarmouth, Norfolk, UK

Printed on acid-free paper

Contents

Preface

By now everyone has heard of the current ecological crisis. Some see it as the main challenge to modern civilization. There are other challenges as well: for example, the wave of global terrorism unleashed by the 9/11 attacks in New York, overpopulation, war, poverty, etc. Without trying to trivialize the significance of problems outside the immediate sphere of their competence (ecology), the authors of this book would like to propose a systemic ecological approach that offers an integrated view of the critical state in which modern civilization finds itself. The aim is new insight into the profound invisible links between seemingly unrelated surface manifestations of one complex problem.

That the crisis has attained dangerous proportions should be obvious even to the most uninitiated observer. And it may very well be that the upsurge in international terrorism—arguably the most conspicuous phenomenon today—is just one of the many symptoms of the general malaise, along with calamitous changes in the environment, explosive population growth in developing countries, large-scale technologically generated disasters and epidemics involving hitherto unknown infections. And all these issues constitute a threat not merely to individual nations or states but to world civilization, to humanity as a whole. The planetary scale of the challenge before us cannot be ignored—no matter how strong the temptation to hide one's head in the sand.

Generally speaking, the notion of challenge in its philosophical and historical aspects was introduced in the middle of the 20th century by British historian Arnold Toynbee in his famous 12-volume work entitled *A Study of History*. As a Christian religious thinker, Toynbee understood challenge as a constantly renewed dialogue between humanity and Divine Reason (Logos) that reveals to people their true essence and supreme historical purpose. Essentially, every endurance test—be it a challenge from nature or from a hostile human community—is, according to Toynbee, the real engine of the historical process. This is the mechanism that awakens a human community's creative energy and raises it to a new stage of development, occasionally also facilitating the emergence of sub-civilizations. A challenge prompts growth, as Toynbee argues. In response to a challenge, society solves a given problem that it faces and in so doing moves to a higher and better state characterized by a more sophisticated structure.

And yet, in the infinitely long challenge–response succession going back to the dawn of antiquity—whether this is seen as trials sent from above or stages in natural historical progress—the current environmental challenge occupies a special place. For the first time in the past few millennia a sacramental question has been raised before the human race or humanity as a biological species: "To be or not to be?" And we are not talking about the threat of a collision between the Earth and an asteroid—an extremely unlikely event that may occur once in many thousands or even millions of years—but rather about the most mundane process of environment degradation and destruction through man's economic activity that has reached a critical limit in the course of its headlong acceleration.

To what extent are the more than 6 billion humans living on the planet aware of this? Not counting professional ecologists, how much is truly known by the proverbial "man in the street" or even the majority of politicians and a substantial proportion of the cultural and business elite? To be sure, such activists as Al Gore have awakened the globe's slumbering ecological consciousness to a considerable extent. However, is the current level of knowledge enough to translate "understanding" into action? It is the answer to this question that will largely determine the fate of future generations.

The truth is that people know both much and little at the same time. "Much" thanks to the media which keep bombarding us with various apocalyptic warnings about the growing greenhouse effect and the resultant widening holes in the ozone layer or the merciless decimation of the tropical rain forest belt known as the lungs of the Earth. "Little" thanks to the surprising ability of our workaday consciousness to retreat from this kind of information into the cozy shade of ideas and myths about the allegedly extreme remoteness of these sinister prophecies or about the possibility of finding shelter behind the secure armor of so-called technological progress.

Given all this, is the ecologists' hope of reaching hearts and minds realistic? After all, their specialized scientific knowledge is accessible, or at any rate comprehensible, to a fairly narrow circle of experts while forecasts and conclusions that follow from this knowledge are directed at people who have nothing to do with ecology as a basic science. However, it is these non-specialists who will ultimately determine the effectiveness or ineffectiveness of a message meant to have a tremendous effect on the destiny of the world.

And yet, it appears that history has already seen precedents or perhaps analogues of a similar situation. In 1939 Albert Einstein, pressed by his émigré colleagues from Western Europe, wrote a letter to the president of the United States, urging Franklin D. Roosevelt to launch a full-scale project for the development of nuclear weapons.

Nothing definite was known about this matter at the time. There was no reliable information beyond guesses and suppositions as to whether Nazi Germany was working on a similar project. Not one physicist could guarantee that the theoretically predicted chain fission reaction of uranium-235 would indeed result in a nuclear explosion. Furthermore, in those years the physicists engaged in researching the atomic nucleus had no influence to speak of and were unknown outside narrow academic circles.

Much was at stake then although the responsibility for making a political decision rested not with scientists but with the president and his team who possessed no specialized knowledge in nuclear physics. Nevertheless, the momentous decision was made after all. Its eventual effect on the political makeup of the world went considerably beyond the limits of World War II, yielding far-reaching positive results in the form of the nuclear deterrent factor, as well as negative ones in the shape of the nuclear arms race and the modern threat of nuclear terrorism.

This virtually "textbook" example is relevant because today's ecological scientists find themselves in a situation somewhat reminiscent of what was faced by the nuclear physicists in the late 1930s. After all, ecologists are also not in a position to produce the "bomb" ticking away under the foundation of modern civilization. Their predictions and estimates are based not on some known precedents, but rather on the logic of biospheric processes that will sooner or later bring (if they have not done so already) irreversible change.

The trouble is that when this irreversibility becomes apparent to the majority, the time needed to make vital decisions will most likely have run out. And only trust in anticipatory scientific knowledge, as was the case with the American nuclear project, can serve as a more or less reliable basis for the prevention of the ecological disaster in question. In this sense the scientists' responsibility, so often mentioned in different respects, is today heavier than ever before. This applies to the responsibility of the world's political, cultural, and business elite, as well as that of all those commonly referred to as people of goodwill. There is probably no task more pressing than for all these human beings to take a significant step toward one another. Hopefully, this book too will do its part in making a contribution to the solution of the urgent problem at hand.

This book is divided into six parts.

Part I: Civilization in crisis: On the edge of an abyss begins with an overview of the ecological, demographic, and other aspects of the global crisis. The global environmental situation involves unprecedented atmospheric change. Destruction of natural ecosystems is shown to be the primary cause of the crisis. With the chemicalization of the Earth's biosphere (48 tons of waste per capita) the global cost of local environmental cleanup is huge. The planet is critically overpopulated. Demographic growth is discussed in the light of biological constraints on the numbers of species. The energy and resource components of demographic growth are explored. Anthropogenic pressure is viewed through the prism of human ecological equivalents.

Part II: Civilization teetering over the abyss of crisis (conclusion) brings Part I to a conclusion. The social dimensions of the crisis are explored, and it is concluded that poverty is a cause of pollution. The problems of hunger and malnutrition, and the harmful environment of the megalopolis are discussed. The "contribution" of the centralized economy and the pros and cons of the market economy are discussed, as well as the market's unreceptiveness to long-term strategies. The limitations placed on the free market by the environment are derived. Ultimately, it is the spiritual crisis of man that is the primary cause of the challenge to the environment.

Part III: The world community: Politicians and scientists in search of a solution provides a history of the first ecology forums and the establishment of a systemic approach to the study of the biosphere. The theory of biotic regulation and stabilization of the environment is the next step in the development of a systemic approach. This part reviews efforts by the Club of Rome, and programs of change from Stockholm to Rio to Johannesburg. It suggests approaches toward a systemic understanding of the biosphere and the concept of biotic regulation as a theoretical basis for sustainability. In this discussion, the following topics are covered: (1) the universal role of the biota in the maintenance of chemical and physical parameters suitable for life, (2) the biotic mechanisms of compensating for disruptions, (3) the destruction of the biotic regulation mechanism under the influence of anthropogenic pressure, and (4) the conservation of ecosystems still in existence as well as the partial reconstruction of destroyed ecosystems as humanity's key goal.

Part IV: Sustainable development: Between complacency and reality deals with the basis of sustainability in nature and in civilization. Two concepts are interpreted: (1) sustainability and sustainable development, and (2) principles of evolutionary strategy in nature vs. growth and development in society. National programs aimed at moving toward sustainable development are inadequate. Finally, the benefits of the nature–society co-evolution concept are discussed. In the process, the evolution of the biosphere "toward humanity" is contrasted with the evolution of humanity "toward the biosphere", as the only alternative.

Part V: On the scale of a scientific approach deals with sustainable development in the context of the biosphere's carrying capacity. Among the topics covered are: (1) distribution of energy streams in biota among groups of organisms, (2) energy quota of man as a large mammal, and (3) the law of energy stream distribution and the biosphere's carrying capacity. In addition, the starting conditions of sustainable development and the safety of ecosystems are given by country and continent. Specific topics include: (a) world centers of environmental stabilization and destabilization, (b) absorption of anthropogenic carbon by the World Ocean and ecosystems on dry land, (c) the rebirth of destroyed ecosystems: a path toward the stabilization of atmospheric carbon, and (d) transition to sustainable development and starting conditions in various countries of the world with respect to the degree of natural ecosystem preservation. Finally, sailing directions and the compass are provided: indicators of sustainable development.

Part VI: "Is there enough community, responsibility, discipline and love?" is the final part of this book. It deals with the barricades of backward thinking and dwindling chances for the future as well as psychological obstacles on the way to grasping the ecological threat. It summarizes what the market economy can and cannot do, and outlines the creative and destructive potential of the market economy. It discusses sustainable development and the actual condition of man. Sustainable development is put forward as a world idea. The social prerequisites of sustainable development and the relationship to globalization are summarized.

Acknowledgments

This section normally opens every monograph. However, chronologically speaking, it constitutes the completion stage which gives the author (or authors) a welcome opportunity to remember all those who have contributed to making this completion possible—from backstage. As one of the authors and the project's coordinator who directed the entire undertaking of preparing the Russian original for publication in English, I would like to give their due to individuals without whose collaboration the monograph would either not have seen the light of day or would have been considerably inferior to its present shape.

I wish to thank the following professors from the Faculty of Biological Sciences at Moscow State University: T. G Korzhenevskaya whose selfless help and moral support were immeasurably important for me at the earliest stages of this long journey; D. V. Karelin whose detailed analysis (20 pages long) of the manuscript's first draft will forever remain in my mind as an example of corporate solidarity and collegiality; and N. N. Marfenin, whose book *The Sustainable Development of Humanity* (so kindly sent to me by the author at the most critical stage of my work) offered invaluable assistance as a kind of encyclopedia on sustainable development.

I will never forget my debt to Professor S. Polozov from Concordia University for generously sharing his writing experience with me and providing me with navigational directions in a sea of books as I was still searching for a publisher. My special thanks go to Professor A. Kaplan from the Johns Hopkins University for his tireless and friendly (and sometimes terse to almost abrasive) critical comments and encouragements. I am also indebted to Dr. M. Bakhnova Cary from the Earth Charter Center on Education for Sustainable Development (Costa Rica) for her goodwill and valuable advice which neither her position nor our strictly electronic relationship obligated her to offer.

Just as one normally does not thank one's fellow authors, it is not customary to thank a book's translator who in a way becomes a co-author by creating something like a new text in the process of transferring ideas from one language to another.

However, in this particular case I would like to express my special gratitude to Dr. Vladimir Tumanov whose creative approach was invaluable to the project. Thanks to the translator's "pre-editorial" efforts, the book was brought more in line with the expectations of readers beyond the realm of Russian culture.

We were all very fortunate to enjoy the services of Dr. Donald Rapp, the outstanding copy editor at Praxis Publishing whose abilities and diligence cannot be underestimated. In particular, I want to stress his great flexibility in the face of changes that were made to the manuscript by the author(s) and translator at different stages. Dr. Rapp's impressive graphic skills made it possible to maximize the communicative potential of the numerous figures and tables in the book. Without his expertise in numerous scientific fields, many a rough spot may have remained ... rough, not to mention his meticulous checking of all the calculations. In short, Dr. Rapp's contribution was just like that of the translator—a labor of love.

Only I will be in a position to fully appreciate the contribution made by my son Dr. Vitaly Reyf. It was specifically thanks to his advice that I was able to include much material that has enriched the book and made it so much more vivid. As a specialist in the biological sciences, Vitaly was often my first counsellor and reliable mediator in my correspondence with the publisher, making it possible to overcome the language barrier.

Every effort of this sort is usually associated with a lengthy period of strain and stress. I cannot imagine how I would have been able to cope with the marathon-like process without the help of my wife Zaytuna Aretkulova. In the one and a half years that it took to essentially create a new book out of the Russian original, she "manned the rear" and patiently put up with my frayed nerves—alas, an inevitable side-effect of such work.

Finally, I would like to express my heart-felt gratitude to our publisher Mr. Clive Horwood, as well as our reviewers, who went over the manuscript samples at a time when the book was far from its final form. By putting their faith in our abilities and giving the project the green light, they offered us a kind of advance which has hopefully paid off as much as possible.

Igor E. Reyf

Figures

Tables

Abbreviations

CFC	Chlorofluorocarbon
CSD	Commission on Sustainable Development
DDT	Dichlorodiphenyltrichloroethane
GDP	Gross Domestic Product
GIS	Geographic Information System
HDI	Human Development Index
ISEW	Index of Sustainable Economic Well-being
IPCC	Intergovernmental Panel on Climate Change
IBP	International Biological Program
IBRD	International Bank for Reconstruction and Development
IBRCE	International Birding and Research Centre—Eilat
IMF	International Monetary Fund (IMF)
MAC	Maximum Allowable Concentration
MIT	Massachusetts Institute of Technology
NGO	Non-Government Organization
NIC	National Intelligence Council
NPP	Net Primary Production
OESC	Office for Economic and Social Council (U.N.)
SARS	Severe Acute Respiratory Syndrome
UNCED	United Nations Conference on Environment and Development
UNESCO	United Nations Educational, Scientific, and Cultural Organization
UCS	Union of Concerned Scientists
UNEP	United Nations Environment Program
WCED	World Commission on Environment and Development
WTO	World Trade Organization
WWF	World Wildlife Fund for Nature
WWII	World War II

Part I

Civilization in crisis: On the edge of an abyss

Part I

1

Global ecological situation

Unprecedented atmospheric change—Destruction of natural ecosystems as the primary cause of the crisis—Chemicalization of the Earth's biosphere—Forty-eight tons of waste per capita—Global cost of local environmental cleanup

It would seem that the global ecological problem has gradually taken on the importance it deserves in the consciousness of the world community. This is evident at the very least from the decision to award the Nobel Peace Prize for 2007 to the former Vice President of the United States Al Gore and the Intergovernmental Panel on Climate Change (IPCC) "for their efforts to build up and disseminate greater knowledge about man-made climate change, and to lay the foundations for the measures that are needed to counteract such change" (IPCC: *http://www.ipcc.ch/*). Despite this progress, questions remain regarding the acuteness of society's ecological awareness. For example, in *Global Trends 2015*, a report of the National Intelligence Council, the word "crisis" appears only in the plural: regional crises, financial crises, etc. There is no mention whatsoever of a global crisis, which is in a way indicative of the problem at hand (*Global Trends 2015*, 2000).

Nevertheless, the growing Green movement, the rising number of agencies for the preservation of the environment in various countries, as well as the growing momentum of the environmental protection technology market, make it clear that the ecological threat has finally been recognized by humanity (e.g., Hansen, 2004, 2005). The shift in our mentality is taking place at a historically unprecedented pace. "Thirty years ago," as V. P. Dolnik notes, "the looming ecological catastrophe and demographic collapse were the concern of only a few ecologists on the planet—these specialists were branded as *alarmists* and mocked from every direction—whereas today, large numbers of ordinary people have become aware of growing pressure from primary factors"[1] (Dolnik, 1992).[2]

Indeed, it would be difficult today to find a country with a population indifferent to the ecological menace, both at regional and local levels ("Global warming", *New York Times*, January 21, 2008). This topic often dominates newsprint and television reports. We keep witnessing the appearance of new periodicals devoted to environmental issues. Not one electoral platform can now avoid promises to tackle this or that ecological problem. New local branches of the world-famous Greenpeace organization keep popping up like mushrooms. And the Green movement has found representation not only in various parliaments but also in governments (e.g., in the German Red–Green coalition of 1999–2005), thereby exerting direct influence on State policies including investment in environmental protection projects.

Although it might appear that all the necessary technological and financial resources have been directed at the current ecological crisis, the problem keeps blocking the path of world civilization like an enormous iceberg and shows no signs of "melting away". The majority of the Earth's population has gradually accepted the notion that the "ecological question" is here to stay and it will occupy their children as well as their children's children with no apparent possibility of returning to the relatively comfortable well-being of the recent past.

And, indeed, the current ecological situation is radically different from all that humanity has ever confronted in its history. This is due to the fact that the dangerous transformation of the environment has become global. These changes have affected all the subsystems and components of the environment all over the surface of the planet, right down to its poles, leaving untouched perhaps only the ocean depths, as has been confirmed by various scientific studies.

Particularly indicative in this connection are the atmospheric dynamics of the changes in the concentration of biogens, substances that participate in life-sustaining processes. Studies of air bubbles from Antarctic and Greenland glacier core samples, which constitute an atmospheric record from the remote past, indicate that the current rate of change in biogen concentration has been unmatched in the Earth's atmosphere for at least 160,000 years (Barnola *et al.*, 1991; Cannariato *et al.*, 1999).

In the past 160,000 years, an increase in biogen concentration similar to that experienced recently has occurred only twice in the Mickulinsky (Russia), Riss-Wurm (Alps), and Sangamonian (North America) interglacial stage, and the modern interglacial period (Holocene). However, the rate of change during these increases was two orders of magnitude less than the speed of today's pace, namely, the changes took place over about 10,000 years.

Every schoolchild probably knows that greenhouse gases (water, carbon dioxide, methane, etc.) play an important role in the so-called greenhouse effect. However, it is less well known that this effect is no less essential for maintaining life on Earth than the planet's atmosphere itself. Greenhouse gases intercept part of the longwave radiation emitted by the Earth, thereby warming up the lower atmospheric strata.

---

[1] These are factors that directly limit the vital activity of the human species.
[2] Inserting "ecological catastrophe" into "Google" leads to 137,000 web pages.

The result is a greenhouse gas generated increase in surface temperature of approximately 40°C compared with the frigid snowball Earth we would have if there were no greenhouse effect. Therefore, it is not the greenhouse effect as such that constitutes a source of danger, but rather the increase in the greenhouse gas that may cause additional warming ("too much of a good thing").

Current perceptions regarding CO_2 and global warming are based on blends of data and suppositions. We start with the data. The CO_2 content has been monitored with instruments since about 1958 and it has risen from 315 ppm in 1958 to about 383 ppm in 2008. Ice core records indicate that over the past four ice age cycles (about 400,000 years) the CO_2 concentration varied between about 190 ppm during extreme glaciation and about 280 ppm during interglacial periods. It is likely that the CO_2 concentration was near 280 ppm in the mid-19th century (Rapp, 2008). Hence the sudden rise in CO_2 in the 20th century is clearly anomalous and is almost surely due to emissions from human economic activity (burning fossil fuels and land clearing). Of the total carbon dioxide emitted into the atmosphere through human economic activity from the start of civilization (according to some estimates 360 billion tons), the overwhelming amount corresponds to the last one hundred years. And the speed of this process keeps increasing inexorably. From 1950 to 1996, yearly carbon emissions from industrial sources have grown 4.6 times (*Global Environment Outlook 2000*, 1999). The year 1996 was a record-breaker with 6.52 billion tons of carbon (the mass of CO_2 is recalculated as carbon by means of a 3.664 coefficient). After the year 2000, carbon emissions stemming from human activity approached 7 billion tons per year, and is estimated to be nearly 9 billion tons per year in 2008. The rapidly evolving economies of China, India, and Brazil, as well as the growth in the number of the world's motor vehicles, has driven this figure rapidly upward (Watson *et al.*, 2001; World Bank, 2007).

Land utilization is also important in releasing CO_2 into the atmosphere, contributing 180 billion tons of atmospheric carbon from the beginning of the Neolithic Revolution until today. In comparison, industrial CO_2 emissions by 2008 constituted roughly 300 billion tons of carbon according to some estimates (Lashof and Ahuja, 1990; Titlianova, 1994; de Laat, 2004). One of the main causes in this case is the destruction of natural ecosystems. Here deforestation is especially significant since forests play a key role in the fixation of atmospheric carbon as part of photosynthesis.

Models for CO_2 distribution indicate that the Earth is only absorbing about half of the CO_2 generated by human activity, and the remainder ends up in the atmosphere as an increase in concentration. Thus, the data show that the atmospheric concentration is rising rapidly, mainly because of human activity, and present levels of CO_2 exceed those that prevailed through the past four ice age cycles.

Another relevant set of data is the change in temperature of the Earth over the past hundred years or so. These data indicate that during the 20th century, the average temperature of the Earth warmed by about 0.7°C (Figure 1.1), although it did not do so continuously, and warming was greater in the north than the south. Temperatures near the end of the 20th century have been compared with the temperature histories over the past 1,000 to 2,000 years as determined by proxies (ice

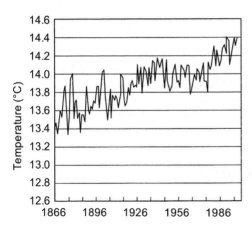

Figure 1.1. Dynamics of surface temperature on Earth (Worldwatch Database, 1998).

cores, tree rings, sediments, etc.) to see how recent fluctuations compare with those from before large-scale human activity. However, a widely quoted study of historical temperatures (Mann *et al.*, 1998, 2003) contains mathematical flaws (McIntyre, 2008; Rapp, 2008) and the comparison of recent temperature fluctuations with historical fluctuations remains unresolved (Jaworowski, 1997, 2007; Kondtratyev, 1999; McIntyre, 2008; Rapp, 2008).

Considering the rate of CO_2 growth in the 20th century and the dependence of the world on fossil fuels, it seems likely that the CO_2 concentration will continue to increase in the 21st century, although limitations on the availability of fossil fuels will gradually reduce the rate of carbon emissions. It is widely believed that the addition of CO_2 to the atmosphere during the 20th century was responsible for the global warming that took place, and that further increases in CO_2 will produce additional global warming during the 21st century (Figure 1.2). A number of groups have developed climate models to predict the amount of such warming based on a benchmark of eventual doubling of the pre-industrial level of 280 ppm to 560 ppm near the end of the 21st century. These models generally predict a modest rise in Earth temperature directly due to CO_2 because the CO_2 absorption bands are saturated and only small increases in thermal absorption occur as the CO_2 concentration is raised. However, when the Earth temperature increases, more water vapor is emitted and water vapor is a very active greenhouse gas, so the models predict a secondary rise from increased water vapor. When the temperature rise from doubling of CO_2 is summed over CO_2 and H_2O, the modelers predict a temperature rise in the 21st century somewhere in the range 1.5°C–4.5°C. However, along with water vapor there are clouds, and clouds are difficult to model; thus uncertainty is introduced in the models (Rapp, 2008). Furthermore, many of those who have projected future increases in CO_2 do not seem to have taken into account limitations on fossil fuel resources. As a result, the world seems to have bifurcated into opposing groups of alarmists and doubters.

If the alarmists prove to be right, severe impacts to the Earth may result. A global surface temperature rise of 1.5°C–4.5°C could precipitate a kind of chain reaction

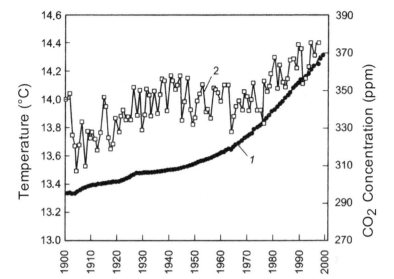

Figure 1.2. Progression of change in the Earth's surface temperature (1) and CO_2 concentration in the Earth's atmosphere (2) (Worldwatch Database, 2000).

stemming from the melting of Arctic ice; the latter would release the CO_2 and methane trapped in permafrost which is also a greenhouse gas (*Climate Change*, 2001; Stott, 2002). The global consequences of such events may include redistribution of the world's climate zones; a rise in World Ocean levels accompanied by the flooding of low-lying coastal regions where almost one-third of the Earth's population lives; and the transformation of the natural environment, threatening the very existence of large segments of humanity.

In general, the destruction or distortion of natural ecosystems (communities of organisms along with their environments: forests, tropics, steppes, tundra-forests, etc.) by human economic activity is without a doubt a most important and essential aspect of the global ecological crisis. The vital and irreplaceable role of these ecosystems in the regulation and stabilization of the biogen cycle will be dealt with more than once in this book. For now, however, it must be stressed that the worst blow to the various ecosystems was delivered in the 20th century.

At the turn of the 20th century, territories with ecosystems completely destroyed by humans took up only 20% of the dry landmass, but by the end of the 20th century it had increased to 63.8%, not counting glaciated and denuded areas. In the northern hemisphere, three enormous zones of environmental destabilization were formed: European, North American, and Southeast Asian, covering a total area of 20 million km^2 (Arsky *et al.*, 1997; Danilov-Danil'yan and Losev, 2000; Nowinsky *et al.*, 2007). However, this process was launched by ancient agriculture.

Several millennia before the industrial era, land "clearing" for agricultural use had already led to the displacement and destruction of tremendous masses of the natural biota (the aggregate of plant and animal organisms). However, this process reached a global scale only at the beginning of the Industrial Revolution. "Hard-working farmers who looked ahead only to a given year's harvest," to quote

L. N. Gumiliov (1993), "ended up transforming the shores of Khotan-Daria and Lake Lobnor[3] into sand dunes; they turned up the soil of the Sahara, thereby allowing sand storms to blow it all away." The forests were among the first to suffer from slash-and-burn agriculture that seemed to be the simplest form of land clearing. Abandoning depleted soils, ancient farmers kept moving to new territories, which at that time appeared to be inexhaustible. (cf. Kalnay and Cai, 2003).

In the pre-industrial period alone, according to various estimates, 30%–50% of forests were destroyed all over the Earth. Another 9%, of which tropical forests occupy first place, were eliminated in the last 200–300 years. Unfortunately, there is no indication that this process has slowed down in any way today. The surface area covered by natural forests keeps shrinking by approximately 1% per year. Furthermore, most of the forests in the industrialized world have undergone radical structural changes. Essentially, the term "forest" in the Western world refers to forestry farms or so-called secondary forests that are at a given stage of natural recovery after stumping, timber felling, or fire damage. On the other hand, European primary or natural forests (not counting Russia) have survived only in such mountainous areas as the Alps, the Pyrenees, the Carpathians, and the Balkan Peninsula, as well as Northern Scandinavia, with a total area of only 450,000 ha (*Protecting the Tropical Forests*, 1990).

Forest ecosystems constitute a vital element of the mechanism responsible for the formation and stabilization of the environment. Accumulating and evaporating water, they facilitate the most important part of continental moisture circulation, maintain the stability of river runoff, slow down the movement of air near the ground (which reduces meteorological extremes), and act like filters countering atmospheric pollution.

Finally, most photosynthetic production takes place within forests. *Net primary production* is a term referring to the part of photosynthetic production that is not used for respiration and growth by the plants themselves, and can be utilized by other organisms (e.g., bacteria, mushrooms, and animals). If one compares the initial productivity of a virgin forest with the productivity of secondary forests that replace them, it turns out that such a substitution equals the loss of approximately 11.7% of net primary production per hectare of forest.

In terms of their biomass production, secondary forests are even more inferior to primary ones. For example, the amount of biomass on one hectare in secondary European forests is half the amount of biomass that was there before forest reclamation. That difference jumps to a whole order of magnitude in artificial forest agrobiocenosis areas. In comparison with natural ecosystems, the total loss in destroyed ecosystems dominated by humans is up to 27% of net primary production (Vitousek *et al.*, 1986; Betts, 2006).

No less harmful for the humus layer is agricultural tillage, along with soil compaction by means of farming technology, leading to its gradual degradation. In the absence of appropriate agronomic methods the humus layer is destroyed

---

[3] An intermittent river and an undrained lake, respectively, in Western China (Sinkiang Uighur Autonomous Region).

completely. Exemplifying the kind of irreparable damage sustained by the humus layer through thoughtless agricultural techniques is the reclamation of virgin land in Kazakhstan and the Altai region. The ecological consequences of this process in the 1950s were very severe: mass degradation of land, water and wind erosion, as well as formation of dust storms. All told, in the entire world, soil mantle loss from erosion today amounts to 6 million ha per year.

However, agriculture is not solely responsible for soil erosion and degradation. Soil is a key component in the biogeochemical cycle. It is the medium in which water accumulates over open dry land and therefore constitutes a kind of "land ocean" providing moisture for the vegetational biota and maintaining the continental moisture cycle. On the other hand, soil is also the habitat for a great variety of decomposer organisms: mushrooms, bacteria, and invertebrates. Thus, in one square kilometer of soil (30 cm deep) there are more than a trillion microorganisms and mushroom filaments that return dead organic elements to the environment. Such elements are biogens whose supplies are limited in nature. Unfortunately, these are the very organisms that die first as a result of agricultural tillage and application of mineral fertilizers and pesticides. Thus, according to data from ecological studies, the introduction of $3 \, \text{g/m}^2$ of nitrogen into the soil decreases the numbers of its resident species by 20%–50% (Vitousek, 1994).

Given that agricultural areas occupy approximately 30% of the total dry land-mass, of which 10% is made up of arable land, and given that the disconnectedness of the biogeochemical cycle[4] on such fields reaches tens of percent, one can easily imagine the scale of the destructive disruption of the biospheric balance by modern agriculture. To this, we should add the hundreds of thousands of hectares that are neutralized every year through the salinization and erosion of soils, which amounts to millions of hectares of land no longer available for the Earth's biosphere.

The large-scale destruction of natural ecosystems also influences the continental moisture cycle, 70% of which is controlled by the vegetational biota. Deforestation, which began as early as the Neolithic Revolution, has played a significant role in the expansion of arid areas, whose extent has now reached 41% of the world's dry landmass. These regions are home to more than 1 billion people, and further desertification is a disaster in the making for the affected populations. This was already demonstrated by a drought and famine in the 1970s in the Sahel zone (south of the Sahara), as well as in Eastern and Southern Africa in the 1990s.

The rate of desertification is also increased through anthropogenic pressure on semiarid ecosystems, especially as a result of massive water diversion for irrigation. In a hot climate, this can lead to quick soil salinization; a clear example thereof is the ecological disaster of the Aral Region. As a result of nearly total water diversion from the Amudaria River for cotton crop irrigation, a process which went on for many

---

[4] This is the synthesis and decomposition of organic matter. The precise interconnectedness of this process in natural ecosystems ensures that the latter remain stable for thousands of years. In artificial agrobiocenoses the removal of biogens along with the harvest by humans leads to soil depletion and the need to maintain soil productivity through the introduction of the required biogens along with fertilizer.

years, the water level of the Aral Sea has dropped since the early 1960s by more than 20 m, while water salinity has increased by three times. The effect of wind-blown salt dispersion and a drop in the ground water level led to the acute deterioration of the climate, as well as the salinization and degradation of the soil in a huge region with a population of 30 million people.

Although pollution of the aqueous environment, as opposed to the pollution of the atmosphere, does not constitute a global ecological threat because the water mass in the World Ocean is 9–10 orders of magnitude greater than extraneous anthropogenic flow, aqueous pollution of fresh water has already reached continental levels. Pollution has affected most enclosed or semi-enclosed seas (e.g., Caspian, Azov, Baltic, the North Sea, etc.):

> "Mechanized man, having rebuilt the landscape, is not rebuilding the waters. The sober citizen who would never submit his watch or his motor to amateur tampering freely submits his lakes to draining, filling, dredging, pollution, stabilization, mosquito control, algae control, swimmer's itch control, and the planting of any fish able to swim. So also for rivers. We constrict them with levees and dams, and then flush them with dredging, channelizations, and the floods and silt of bad farming" (Leopold, 1941, p. 17).

In this connection we ought to keep in mind that, along with the World Ocean, rivers and lakes in some ways represent the final stage in the continental cycle of pollutants. It is here that fertilizers and pesticides from agricultural land end up, along with discharges from industrial plants and urban areas. Finally, sooner or later, atmospheric pollution drops down and is carried by melt or rain water to the surface of river catchment areas. Therefore, one should not be surprised that in the bottom sediments of certain polluted water bodies, one can find nearly the entire periodic table.

Most of the developed world should also heed a warning about fresh water. Many industrialized countries have almost pushed their water consumption to its uppermost limit, and in some states (e.g., Belgium), water intake has reached 70% of all renewable water resources. Despite the huge expenditure of funds on water-purifying installations, the quality of surface water in Europe remains very low. The Elbe, the Oder, the Dnieper, the Southern Bug, and the Guadalquivir can all be classified as extremely polluted rivers according to accepted parameters. They have high levels of pesticides and other dangerous organic compounds. As for the concentration of certain metals (lead, chrome, zinc, etc.), the waters of the Elbe, for instance, contain 3–16 times more than the background (*Europe's Environment*, 1995; *World's Worst Polluters*, 2006).

After 1940 the process of anthropogenic eutrophication or water bloom, the vigorous proliferation of certain algae species under the influence of biogenic elements accumulating in surface waters, reached significant proportions. Toxins formed as a result of water bloom, as did the oxygen deficiency characterizing such water (caused by the rapidly multiplying oxygen-consuming anaerobic bacteria which feed on dead organic matter), bringing about the mass deaths of bottom organisms (Green *et al.*, 1984).

In nature, eutrophication takes place under natural conditions. But, in nature, this process takes thousands of years and its speed cannot compare with the rate of anthropogenic water bloom. The latter is provoked by the outwash of nitrogen-based fertilizers from agricultural land and the flushing of sewage rich in phosphorous-containing compounds into water reservoirs, primarily from highly urbanized areas.

At the same time the increased scale of water regulation (river canalization and bunding, as well as the construction of dams and water reservoirs, etc.) is undermining the self-cleaning capacity of river water. A typical example is the Volga cascade that has turned Russia's main artery into a chain of gigantic water reservoirs with significantly slowed-down current and the associated intensive development of eutrophication processes (Paerl, 1997).

Equally significant in the process of water environment degradation is the role that has been played by acidification and salinization of freshwater reservoirs. The direct cause of acidification is acid rain brought about by the emission into the atmosphere of sulfur and nitrogen oxides formed through the process of fossil fuel combustion. As they become absorbed by raindrops, these compounds settle on water and soil surfaces, often poisoning all living organisms. In any case, shrinking forests and small dead lakes with no fish or plankton first appeared in older industrial regions of the U.S., Europe, and Japan in the middle of the previous century. This phenomenon, which has become common since the 1970s, is first and foremost the consequence of acid rainfall.

As for salinization, which was known even in ancient Babylon and Assyria, this process turned into a veritable plague for irrigation farming in the 20th century. Today, nearly a thousand tons of water are required in order to produce one ton of rice. We now witness the use of up to 80% of surface and ground water in the agriculture of rice-producing countries. The water level is reduced catastrophically, which leads to the salinization of fresh water reservoirs. Thus, in the agricultural regions of Northern China, the ground water level keeps dropping by 5 feet (1.5 m) per year while in India it may drop by 3 ft–10 ft per year. And since the role of underground water sources has grown rapidly in the recent past as a result of surface water pollution (in some countries underground water accounts for up to 50% of total intake), the depletion of underground water-bearing strata could lead to palpable fresh water shortages in certain world regions as soon as the next 10–15 years (Brown *et al.*, 2000).

Furthermore, the rising need for fresh water is outstripping population growth. And so, in order to satisfy the increasing demand for food, the segment of the world's harvest from irrigated land will have to be raised from 28% to 46% by the year 2025. However, the planet is already suffering from a fresh water deficit equaling roughly two yearly water flow volumes of the Nile (over 160 million km^3 of water per year). According to estimates from the International Water Management Institute, by 2025 a billion people will be living in countries with an absolute water deficit (less than 1,700 m^3 per person per year). These would include parts of the Near East and South Asia, almost all of Africa, as well as the North of China. As a result there will be even

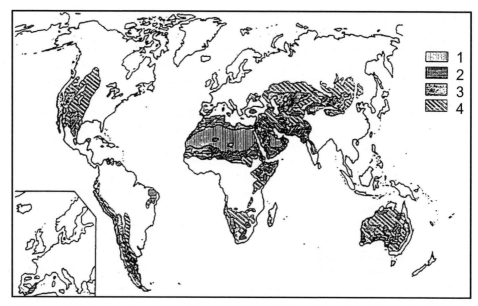

Figure 1.3. Territories subject to desertification. 1: Deserts. 2–4: Territories subject to desertification by degree of danger; 2: very high; 3: high; 4: moderate (Rodionova, 1995).

more acute water shortages for agricultural needs in these regions. And even given more effective irrigation methods, the areas concerned will be unable to maintain their per capita production of food on irrigated land at the level of 1990 or supply water for their industrial, household, and ecological needs. In general, the authors of *Beyond Malthus: Nineteen Dimensions of the Population Challenge* view the growing fresh water deficit as perhaps today's most underestimated resource problem (Brown *et al.*, 1999, pp. 37–39).

Almost the entire dry landmass of the planet, from the circumpolar tundra to the searing sands of the deserts, is covered by a continuous layer of life, interrupted neither by high-altitude plateaus nor by the craters of extinguished volcanoes. This living stratum is the result of long evolution, in the course of which various species and their communities have come to inhabit the entire geoclimatic range of conditions, thanks to the great differentiation of life forms and their combinations. This is known as *biodiversity*, a term familiar today even to the wider public. This is the mechanism that allows every living organism to be maximally efficient in the utilization of natural resources within its habitat and biological niche. The latter can be viewed as the "profession" of each living thing.

Even if life on Earth has suffered periodically from local interruptions caused by catastrophic movements of the planet's crust, volcanic activity, or collisions with asteroids, there have always been life forms capable of surviving these crises and gradually filling the new gaps. This continuous development of life in time is also attributable to the planet's biodiversity, a key factor in sustaining the functional structure of the biosphere and the effectiveness of biogenic processes in ecosystems.

However, with the beginning of large-scale human economic undertakings, this precious evolutionary gain has come under threat. The destruction of natural ecosystems and the technologically mediated transformation of the landscape is undermining the basis for existence of numerous species, as well as their communities, some of which have already vanished from the face of the Earth, while others teeter on the brink of extinction. The situation is made all the more complex because many species are disappearing without being recognized, a problem typified by multitudes of insects and protozoans that live under the tropical forest canopy.

In such cases, scientists have to rely only on estimated data, according to which the rate of biodiversity loss amounts on average to at least one species per day. And if we limit ourselves to the vertebrates, then since the year 1600 the Earth has lost 23 species of fish, 2 species of amphibians, 113 bird species, and 83 mammalian species (McNeely, 1992). Each extinct species constitutes a final and irreparable loss for the biosphere (there is no way back in evolution), and a far greater number of species is in danger of extinction. It is not difficult to imagine that if current dangerous trends do not change, in a hundred years or so the so-called "masters of the planet" will inhabit a barren place, a species desert (Thomas *et al.*, 2004).

As was mentioned above, global warming has been theorized to be linked to the growth of the greenhouse effect brought about by the accumulation of man-made atmospheric carbon dioxide. However, the large-scale interference by humans in the biospheric biogen cycle can also affect the concentration of another greenhouse gas: namely, water vapor. Its proportion in the atmosphere (approximately 0.3%) is an order of magnitude greater than that of CO_2, which accounts for water vapor's significantly greater role in the final greenhouse effect. In this connection, the destruction of natural ecosystems, leading to the disruption of the global moisture cycle, is one more probable cause of destabilization in the Earth's climate.

Any rise of surface temperature that occurs as a result of growing CO_2 concentration would cause an increase in the evaporation from the World Ocean's surface, and therefore the accumulation of excess water vapor in the atmosphere. This vapor in turn would exacerbate the greenhouse effect; that is, global warming, although it is not clear how cloud formation would affect the heat balance. These processes, which are still insufficiently studied but very likely play a major role in the way humanity is changing the climate, may be more significant than the growing concentration of the other greenhouse gases. What remains certain is that the globally disrupted mechanism of the biogen cycle, which is maintained by the natural biota with great precision, is one of the leading factors contributing to destructive climate change.

Along the same lines, some researchers argue that we should consider the possibility of more frequent climate-related natural disasters. The argument is that the Earth's climate is entering a stage of great instability, which may be inferred from the recent series of natural cataclysms. Indeed, natural disasters, along with technological accidents, can (for example) cause leaks from containers and tanks at chemical plants or knock out electrical networks, etc. (see Figure 1.4).

According to an assessment by the Geoscience Research Group, the number of climate-based natural disasters in 1997–1999 grew by a quarter in comparison with

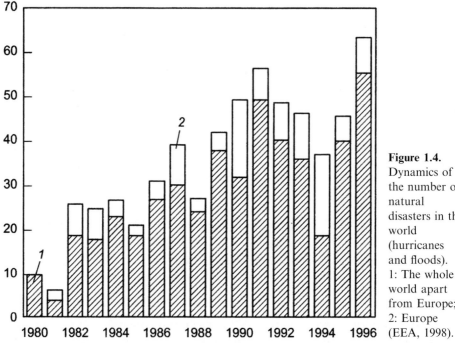

Figure 1.4. Dynamics of the number of natural disasters in the world (hurricanes and floods). 1: The whole world apart from Europe; 2: Europe (EEA, 1998).

the beginning of the same decade. These events have claimed tens of thousands of lives, and the material losses are valued at tens of billions of dollars (e.g., $100 billion in 1999 alone). A study conducted by the American insurance company Travelers Corporation (insurers are the first to bear the brunt of tornadoes and hurricanes) hypothesizes that a rise in average surface temperature of only 0.9°C by 2010 would be sufficient to increase the number of hurricanes smashing into the American coast-line by a third (Chirkov, 2002; van Aalst, 2006). Figure 1.5 illustrates the growth in damage caused by natural disasters in the second half of the 20th century according to data provided by the German insurance company Munich Re (Global Environment Outlook 2003. London: Earthscan Publishing, 2000).

The 2001 IPCC Report required over 1,000 pages to describe the negative impacts of global warming including an increase in violent storms. However, as in the case of the putative contribution of CO_2 to global warming, there are counter-arguments. Webster (2005) found a doubling of the number of Category 4 and 5 hurricanes in the 15-year period 1990–2004, as compared with 1975–1989. However, this article pointed out that they

"... deliberately limited this study to the satellite era because of the known biases before this period, which means that a comprehensive analysis of longer-period oscillations and trends has not been attempted. There is evidence of a minimum of

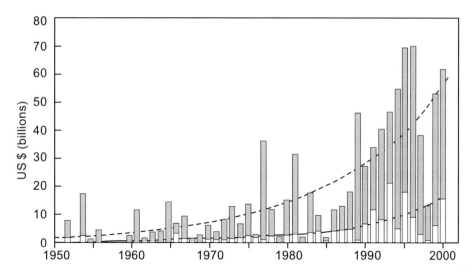

Figure 1.5. Damage from natural hazards, and insurance payments. (Gray columns and dashed line) Natural hazard. (White columns and solid line) Insurance payments.

intense cyclones occurring in the 1970s, which could indicate that our observed trend toward more intense cyclones is a reflection of a long-period oscillation."

Klotzbach (2006) found only a 10% growth in global Category 4 and 5 hurricanes from 1986–1995 to 1996–2005, of which most were in the southern hemisphere. In another publication, Klotzbach said:

"These findings indicate that there has been very little trend in global tropical cyclone activity over the past twenty years, and therefore, that a large portion of the dramatic increasing trend found by Webster and Emanuel is likely due to the diminished quality of the datasets before the middle 1980s. One would expect that if the results of Webster *et al.* and Emanuel were accurate reflections of what is going on in the climate system, then a similar trend would be found over the past twenty years, especially since SSTs have warmed considerably (about 0.2°C– 0.4°C) during this time period."

Robinson *et al.* (2007) presents data that show that the number of tornados in the United States has been on a downward trend from 1950 to 2005. They also show that there has been no statistical increase in the annual number of Atlantic hurricanes that make landfall from 1900 to 2005. The period after 1998 has been relatively high, but there is no evidence yet that this is any more than a fluctuation of the same magnitude as has occurred previously. Similarly, the annual number of violent Atlantic hurricanes shows no statistical trend from 1945 to 2005, although the period since 1995 has been above average. It is impossible to tell whether this is just a fluctuation of the same magnitude as has occurred previously, or whether it is the early emergence of a new trend.

These issues require further study.

A special aspect of the global ecological crisis is the rapid accumulation in the environment of waste from human economic activity, including the products of chemical synthesis with pronounced toxic characteristics. Most of this waste, especially its solid forms, is created through the inexorably growing process of mineral raw material production. By the beginning of the 19th century, no fewer than 300 billion tons of mineral raw materials were being produced and transported every year. This number includes waste created in the process of overburden[5] operations, through construction work, as well as from agricultural activity, as a result of farmland erosion (Arsky et al., 1997).

It has been suggested that waste and environmental pollution constitute the main threat to modern civilization; however, to what extent this is accurate will be discussed below. As for the extent of this damage, the volume of waste generated by economic activity is indeed truly daunting and defies the imagination. Thus, for every inhabitant of the Earth, 50 tons of raw materials are extracted and transported from the bowels of the planet per year. However, only 2 tons of that amount are turned into finished products.

Therefore, having made this gigantic effort, humanity ends up with almost the same amount of waste (48 tons per person) of which 0.1 tons per person is dangerous. In the developing world that figure may even reach 0.5 tons per person (Arsky et al., 1997; Danilov-Danil'yan and Losev, 2000). But even the above-mentioned 2 tons of final product are essentially a form of waste generation, merely postponed into the future as a "present" for subsequent generations. Therefore, from an ecological perspective, almost everything produced by humanity in the material sphere is sooner or later bound to become waste. From this it would follow that even the Egyptian pyramids or archeological cultural strata are a form of very old waste which allows humanity to learn about its history.

Naturally different forms of waste make very different contributions to environmental contamination. In this respect, chemically active substances and products appear to be without equal. Some of them are highly stable and take a long time to decompose, which means that they are preserved and accumulate in all environments, including the human body. Others are destroyed by biological processes and begin to accumulate only when the stream of such substances exceeds the biological capacity for their elimination (Odum, 1971). Short-lived pollutants (with a lifetime not exceeding a week) cause regional contamination when they enter the atmosphere. A lifetime of over 6 months is associated with global contamination.

Aerosols, minute suspended particles with a diameter of 0.1 to hundreds of microns, are typical atmospheric contaminants. *Europe's Environment: Statistical Compendium for the Dobris Assessment* (1995) features a map illustrating the average

---

[5] Material overlying a deposit of useful geological materials or bedrock (*Merriam Webster* [*http://www.m-w.com/dictionary*]).

yearly concentration of atmospheric aerosols over Europe for the year 1992. Here we can clearly see the way an uninterrupted cloud of thin, suspended, industrially generated aerosols with a density exceeding $20 \, \mu g/m^3$ covers huge territories of Central and Eastern Europe, southwestern England, the Benelux countries, and northeastern France.

Aerosols contain both solid (dust, ashes, soot) and liquid components. The latter include sulfur and nitrogen oxides, ammonia, as well as airborne organic hydrocarbons. Furthermore, they absorb many metals, lead in particular, and high-molecular toxic compounds. When they reach human respiratory channels, some of them (dust particles, nitrogen dioxide, sulfur dioxide) act as direct irritants and allergens. Other compounds penetrate into the bloodstream and can provoke a generally toxic effect. In this connection the so-called photochemical smog is especially dangerous because under the influence of solar radiation, this hellish mix of exhaust fumes from vehicles and industrial emissions becomes the locus of photochemical reactions with the formation of ethylene, ozone, and other gases.

Dangerous waste and supertoxicants constitute a special category of pollutants, 90% of which come from industrialized countries. And the U.S. is in first place. Thus, according to data from *The World Environment 1972–1992*, 270 million tons of dangerous waste were produced per year in the United States in the early 1990s. The second place in this list belongs to Russia with 107 million tons. Then comes India with 36 million tons. And these numbers have continued to grow in subsequent years, especially in such countries as China, India, and Brazil. However, such data are usually not publicized; the numbers end up either concealed or lowered.

Thanks to the media, many of the compounds from this group have become familiar to the public (e.g., heavy metals and pesticides), as well as related compounds from the aromatic chlorinated hydrocarbon group (e.g., dioxins, biphenyls, furans, etc.). They are all quite stable in the environment and are not subject to easy chemical or biological decomposition since they are foreign to the biota. Therefore, they can persist for tens of years, penetrating every stratum of the environment and joining the progression of food chains which link various organism species. In this manner, for example, dioxins, formed as a byproduct of many technological processes, have been found not only in the atmosphere, water, and soil, but also in food, including human and animal mothers' milk. The truly global extent of these pollutants is evident from their significant presence even beyond the Arctic Circle. Some of them affect the hormonal, neural, and reproductive systems, hence the term "super-toxicants" (*Okruzhaiuschaia sreda*, 1993; Colborn *et al.*, 1996; Baranowska *et al.*, 2005).

The role of pesticides in the pollution of the soil and water environments is well-known. These substances began their conquest of the planet in 1938 when DDT (dichlorodiphenyltrichloroethane) was discovered by the Swiss chemist P. Müller, whose achievement ended up crowned by the Nobel Prize. And it was immediately after WWII that the mass production of pesticides expanded rapidly. Today there are around 180 pesticide brands around the world, and their total production in the early 1990s amounted to 3.2 million tons (i.e., 0.6 kg for every person on Earth). Admittedly, in the past ten years in the developed world there has been a reduction in pesticide use accompanied by a transition to less dangerous pesticide types. However, in the "Third World" pesticide use is still growing, along with the use of dangerous forms such as DDT.

In ecopathology, pesticides are rated according to the highest stress index, 140 (heavy metals, transported nuclear power plant waste, and solid toxic waste). For every hectare of tillage 0.5 kg–11 kg of these toxic chemicals are used. Half of that amount immediately turns into waste and accumulates as such in the soil as well as ground water. Poor awareness of this problem leads to 500,000 to 2,000,000 pesticide-related accidents all over the world, of which 10,000–40,000 are fatal (*Okruzhaiuschaia sreda*, 1993).

In Rachel Carson's classic *Silent Spring* (1962), one of the first ecological alarm bells, we read that almost the entire human population has become subject to the action of chemicals, and no one can determine the long-term consequences. Today, over four decades later, these consequences are becoming more and more clear. In particular, it turns out that many pesticides, starting with the already banned DDT, as well as polychlorinated biphenyls, dioxins, furans and finally a whole series of metals (cadmium, lead, mercury), are held responsible for disruptions in the human endocrine system, hormonally determined breast cancer and cancer of the prostate, lower sperm counts, infertility, congenital defects, and neurological anomalies in children.

Furthermore, substances in this category are very slow to decompose and there-fore tend to accumulate in the body. Thus, lead concentrates in bone tissue, and as a result its content in the bones of modern humans is almost 1,000 times higher than in the bones of those who lived 1,500 years ago (Khudolei and Mizgiryov, 1996). Chlorinated pesticides and biphenyls accumulate in fat tissue and penetrate into breast milk in fat droplets. Analysis of unpasteurized milk samples demonstrates that even in well-to-do Bavaria every third sample contained biphenyls in con-centrations exceeding the MAC (Maximum Allowable Concentration) (*Okruzhaiuschaia sreda*, 1993). Therefore, the "chemicalization" of the biosphere is a *fait accompli*.

From 100,000 to 200,000 various (including synthesized) substances are in circulation on the world market; for 80% of these we do not know, and are unlikely to ever know completely, how they act upon living organisms. Moving along the food chain, some of these substances can accumulate in its higher links (including man) in concentrations exceeding the baseline by hundreds of thousands and millions of times. We can therefore rightly view our civilization as a giant experimental animal facility where people act as guinea pigs subject to the testing of unknown drugs (Coman *et al.*, 2007).

The question is whether one can still find a way to overcome the boundless sea of chemicals now threatening the very existence of humanity. Can this unending waste be conquered by the newest technology? With respect to the first question, for now there is unfortunately no answer. As for the widespread illusion regarding some sort of technological "magic wand" (which has not yet been created but once it is, all toxic waste and chemicals will be waved out of existence), this notion deserves further attention.

Let us take the burning of trash since this appears to be the most obvious and direct method of solid waste elimination. All one needs to do is expend a certain amount of energy in order to create the required temperature in garbage incinerators. By the way, this tried and tested method is already over 130 years old. However, as of the mid-1980s numerous European and American States authorities have been shutting down such installations. The question is why.

It turns out that, first of all, if solid waste contains both chlorine compounds and variable valence metals, waste incineration causes the formation of highly toxic dioxins. However, the key point is that even though incineration does decrease solid waste volume ten times, the refuse is transformed into a gas whereby every ton of solid waste creates 30 kg of airborne ashes and 6,000 m^3 of fume gases containing sulfur dioxide, nitrogen and carbon oxide, hydrocarbons, heavy metals, as well as the already mentioned dioxins. This whole smoke plume gains direct access into the atmosphere through industrial smokestacks and is then carried by air currents for hundreds and thousands of kilometers. So, which is the lesser of two evils? What harms man and nature more: solid or incinerated waste? Today some countries are returning to incineration, but this time the technology is different: new incineration equipment and techniques, special filters, and the preliminary sorting of waste.

The example of waste incineration merely illustrates the fundamental Lomonosov–Lavoisier law of the conservation of matter according to which there is no way of eliminating waste once it is produced. The refuse can be hidden (buried), transformed from one phase state into another, dispersed in the environment, or finally reprocessed into some other less toxic product. However, the latter will in turn also become waste.

Therefore, the solution to this problem, and unfortunately we are not talking about something radical here, is the creation of resource-preserving technologies or the reorganization of the production process so that one producer's waste turns into raw material for another. The latter scheme has been implemented, for example, in Denmark's remarkable eco-industrial Kalundborg Park. However, such promising technological showpieces conceal a certain amount of non-utilizable waste. Most importantly, Kalundborg's production is itself a form of waste which has merely been put aside beyond the purview of the producer.[6]

In general, waste recycling is used quite widely in the world. The greatest success has been achieved in Japan where 210 million tons (or 10% of the 2.6 billion tons of materials produced yearly in the country) are recycled every year. However, unfortunately all such technologies are expensive, and besides, they require considerable energy expenditure. Any form of energy production inevitably exerts pressure on the environment and in the end causes its distortion and destruction, thereby undermining all positive results.

Turning once again to the world's top recycler (Japan), we see that in 1970–1990 this country was able to restructure its economy in order to sharply decrease the role

[6] Ideally, such systems of production should be subject to rigid internal correlation, but this does not take place in reality. As a result all random changes inside such an entity inevitably cause its disorganization, and therefore environmental pollution.

of the raw material industry and the so-called "dirty" producers. Priority was given to the service sector, information technology, advanced technology, and eco-efficient production based on resource conservation, recycling, and the prolonging of end-product life spans.

The question is how effective all these measures were. It turns out that despite the reduction of the raw material industry, the stream of materials in Japan has not shrunk. In fact, it has grown, thereby causing an increase in the related production of waste. The most important point is that the per capita rate of energy consumption went up by 15% in 1990–1995 (*Quality of the Environment in Japan*, 1999). The situation is similar in the U.S. and in Western Europe. In the last quarter of the 20th century these countries expended huge sums on environmental protection and the transformation of the extensive "dirty" economy into its resource-conserving intensive counterpart. However, it is not by chance that by the early 21st century there has been no significant decrease in per capita energy consumption (*Global Environment Outlook 2000*, 1999). In fact, the case was just the opposite: that is, in many of these countries energy consumption has continued to grow, which, as has been stated above, is a bad ecological symptom (Fischer, 2005).

Neither are there many significant differences in terms of global results between widely publicized measures taken by various states for local environmental cleanup. To be sure, there are individual cases of clear success; for example, the Great Lakes in the U.S.A./Canada and the Rhine in Germany where the level of contamination (especially in the case of the Rhine) was indeed horrendous 40 years ago. However, has anyone tried to calculate the overall ecological balance of this local cleanup effort? For example, how much energy and materials were required, and what were the ecological consequences for those countries that provided these resources? What about the countries to which the technologically "dirty" facilities were transferred?

We ought to keep in mind that an individual ecological success story of a particular country is often paid for, according to the law of communicating vessels, through losses in other parts of the world. And so, total ecological impacts usually exceed advantages from local decontamination. Therefore, the fact that despite improving conditions in certain limited areas, the global ecological situation continues to deteriorate indicates that on a global scale such measures are merely a case of sweeping the proverbial "dirt under the carpet".

In light of the above it is high time to reconsider the famous notion elaborated by the Club of Rome: namely, "think globally and act locally". It is the second part of this phrase in particular that needs reassessment: both thinking and action should be global or at least the results of any local actions and decisions ought to be viewed in global terms.

And so, throughout the entirety of its existence, human civilization has essentially failed to create a single technology which in one way or another does not stress the environment. For many centuries the biosphere kept struggling with some success against the destructive activity of mankind. However, starting in the early 20th century, all environments have displayed unprecedented unidirectional changes whose speed has been growing inexorably. This rate of change is unsurpassed in nature. Therefore, natural regulating mechanisms are no longer capable of withstanding the destructive onslaught of civilization. This ecological crisis has developed within only one generation.

References

Arsky, Yu. M.; V. I. Danilov-Danil'yan; M. Ch. Zalikhanov; K. Ya. Kondratyev; V. M. Kotlyakov; K. S. Losev (1997). *Ekologicheskie problemy: chto proiskhodit, kto vinovat i chto delat'?* Moscow: MNEPU [in Russian].

Baranowska, I.; H. Barchanska; A. Pyrsz (2005). "Distribution of pesticides and heavy metals in trophic chain," *Chemosphere* **60**, 1590–1589.

Barnola, J. M.; P. Pimienta; D. Raynaud; Y. S. Korotkevich (1991). "CO_2 climate relationship as deduced from Vostok Ice Core: A re-examination based on new measurements and on re-evolution of the air dating," *Tellus* **43B**(2), 83–90.

Betts, R. A. (2006). "Forcings and feedbacks by land ecosystem changes on climate change," *J. Phys. IV France* **139**, 119–142.

Brown, L.; G. Gardner; B. Halweil (1999). *Beyond Malthus: Nineteen Dimensions of the Population Challenge*. New York: W. W. Norton & Co.

Brown, L. *et al.* (eds.) (2000). *State of the World 2000*. London: W. W. Norton & Co.

Cannariato, K. G.; J. P. Kennett; R. J. Behl (1999). "Biotic response to late Quaternary rapid climate switches in Santa Barbara Basin: ecological and evolutionary implications," *Geology* **27**(1), January, 63–66.

Carson, R. (1962). *Silent Spring*.

Chirkov, Yu. (2002). "Sindrom utomlenia planety," *Literaturnaia gazeta*, pp. 20–21 [in Russian]..

Climate Change (1990). New York: Melbourne University Press.

Colborn, T.; D. Dumanoski; J. P. Myers (1996). *Our Stolen Future*. New York: Dutton.

Coman, G.; C. Draghici; E. Chirila; M. Sica (2007). "Pollutants effects on human body: Toxicological approach," in *Chemicals as Intentional and Accidental Global Environmental Threats*. Dordrecht, The Netherlands: Springer-Verlag.

Danilov-Danil'yan, V. I.; K. S. Losev (2000). *Ekologicheski vyzov i ustoichivoie razvitie*. Moscow: Progress-traditsia [in Russian].

de Laat, A. T. J.; A. N. Maurellis (2004). "Industrial CO_2 emissions as a proxy for anthro-pogenic influence on lower tropospheric temperature trends," *Geophysical Research Letters* **31**, L05204.

Dolnik, V. T. (1992). "Are there biological mechanisms for regulating human population numbers?" *Priroda* **6**, 3–16. Available at *http://vivovoco.rsl.ru/VV/PAPERS/ECCE/ VV_EH13W.HTM*

EEA (1998). *Europe's Environment: The Second Assessment—Data Pocketbook*. European Environmental Agency.

EPA (Environmental Protection Agency) (2005). Human-related sources and sinks of carbon dioxide. Available at *http://www.epa.gov/climatechange/emissions/co2_human.html*

Europe's Environment: Statistical Compendium for the Dobris Assessment (1995). Luxemburg: Eurostat.

Fischer, G. (2005). "Energy consumption and limits to global emissions of carbon dioxide: Australia and the world," *Population and Environment*, **13**, 183–191.

Green, N. P. O.; G. W. Stout; D. J. Taylor (1984). In: R. Soper (ed.) *Biological Science*, Vol. 1. Cambridge, U.K.: Cambridge University Press.

Global Environment Outlook 2000 (1999). London: Earthscan.

Global Trends 2015 (2000). A dialogue about the future with nongovernment experts. Available at *http://www.dni.gov/nic/NIC_globaltrend2015.html*

Global Warming (2008). *New York Times*, January 21. Available at *http://topics.nytimes.com/top/news/science/topics/globalwarming/index.html*

Gumiliov, L. N. (1993) *Ethanogenesis and the Earth's Biosphere*. Moscow: Rol'f. Available at *http://www.kulichki.com/~gumilev/English/ebe0.htm*

Hansen, J. E. (2004). "Defusing the global warming time bomb," *Scientific American*, pp. 69–77, March.

Hansen, J. E. (2005). "A slippery slope: How much global warming constitutes 'dangerous anthropogenic interference'"? An Editorial Essay. *Climatic Change* **68**, 269–279. Available at *http://www.columbia.edu/~jeh1/hansen_slippery.pdf*

IPCC (Intergovernmental Panel on Climate Change). Available at *http://www.ipcc.ch/*

Jaworowski, Z. (1997). "Another global warming fraud exposed: Ice core data show no carbon dioxide increase," *21st Century Science and Technology* **10**(1), 42–52.

Jaworowski, Z. (2007). *CO_2: The Greatest Scientific Scandal of Our Time*. Available at *http://www.21stcenturysciencetech.com/Articles 2007/20_1-2_CO_2_ Scandal.pdf*

Kalnay, E.; M. Cai (2003). "Impact of urbanization and land-use change on climate," *Nature* **423**, 528–531.

Khudolei, V. V.; I. V. Mizgiryov (1996). *Ekologicheski opasnyie factory*. St. Petersburg: Izdatel'stvo "Bank Petrovsky" [in Russian].

Klotzbach, P. J. (2006). "Trends in global tropical cyclone activity over the past twenty years (1986-2005)," *Geophys. Res. Lett.* **33**, L010805.

Kondratyev, K. Ya.; V. K. Donchenko (1999). *Ekodinamika i geopolitika*, Vol. 1: *Global'nie problemy*. SPB [in Russian].

Lashof, D. A.; D. R. Ahuja (1990). "Relative global warming potentials of greenhouse gas emissions," *Nature* **344**, 529–531.

Leopold, A. (1941), "Lakes in relation to terrestrial life patterns." In: *A Symposium on Hydrology*. Madison, WI: University of Wisconsin Press, pp. 17–22.

Mann, M. E.; R. S. Bradley; M. K. Hughes (1998). "Global-scale temperature patterns and climate forcing over the past six centuries," *Nature* **392**, 779–807.

Mann, M. E.; P. D. Jones (2003). "Global surface temperatures over the past two millennia," *Geophys. Res. Lett.* **30**, 1820.

McIntyre, S. (2008). *Climate Audit*. Available at *http://www.climateaudit.org/*

McNeely, J. A. (1992). "The sinking ark: Pollution and the worldwide loss of biodiversity," *Biodiversity and Conservation* **1**, 2–18.

Munich Re (2000). *Global Environment Outlook 2003*. London: Earthscan.

Nowinski, N. S.; S. E. Trumbore; E. A. G. Schuur; M. C. Mack; G. R. Shaver (2007). "Nutrient addition prompts rapid destabilization of organic matter in an Arctic tundra ecosystem," *Ecosystems*, doi:10.1007/s10021-007-9104-1, 2007. Available at *http://www.springerlink.com/content/t5650v8x5711l87k/*

Odum, Yu. (1975). *Osnovy ekologii*. Moscow: Mir [in Russian].

Okruzhaiuschaia sreda (1993). *Entsiklopedicheskii slovar'-spravochnik*. Moscow: Pangea [in Russian].

Paerl, H. W. (1997). "Coastal eutrophication and harmful algal blooms: Importance of atmospheric deposition and groundwater as 'new' nitrogen and other nutrient sources," *Limnology and Oceanography* **42**(5), Part 2: The ecology and oceanography of harmful algal blooms (July), 1154–1165.

Protecting the Tropical Forests: A High Priority International Task (1990). Bonn: Bonner Universität.

Quality of the Environment in Japan (1999). Tokyo: Institute for Global Environmental Strategies.

Rapp, D. (2008). *Assessing Climate Change*. Chichester, U.K.: Springer/Praxis.

Robinson, A. R.; N. E. Robinson; W. Soon (2007). "Environmental effects of increased atmospheric carbon dioxide," *J. Amer. Physicians Surgeons* **12**, 79–90.

Rodionova, I. A. (1995). *Global Problems of Humanity*, 2nd Edition. Moscow.

Stott, P. A.; J. A. Kettleborough (2002). "Origins and estimates of uncertainty in predictions of 21st century temperature rise," *Nature* **416**, 723–726.

Thomas, C. D.; A. Cameron; R. E. Green; M. Bakkenes; L. J. Beaumont; Y. C. Collingham; B. F. N. Erasmus; M. Ferreira de Siquiera, A. Grainger; L. Hannah *et al.* (2004). "Extinction risk from climate change," *Nature* **427**, 145–148.

Titlyanova, A. A. (1994). "Emissia dioksida ugleroda i metana v atmosferu," *Obozrenie prikladnoi i promyshlennoi matematiki* **6**, 974–978 [in Russian].

van Aalst, M. K. (2006). "The impacts of climate change on the risk of natural disasters," *Disasters* **30**(1), 5–18.

Vitousek, P. M. (1994). "Beyond global warming: Ecology and global change," *Ecology* 7(75), 1861–1876.

Vitousek, P. M.; P. R. Ehrlich; A. H. Ehrlich; P. A. Matson (1986). "Human appropriation of the products of photosynthesis," *Bioscience* **36**, 368–373.

Watson, R. T. *et al.* (2001). *Climate Change 2001: Synthesis Report*. Cambridge, U.K.: Cambridge University Press.

http://www.grida.no/climate/ipcc_tar

Webster, P. J.; G. J. Holland; J. A. Curry; H-R. Chang (2005). "Changes in tropical cyclone number, duration, and intensity in a warming environment," *Science* **309**, 1844–1846.

World Bank (2007). *Policy Research Working Paper 4352: Atmospheric Stabilization of CO$_2$ Emissions*. G. Timilsina.

World Environment 1972–1992. Available at *http://www.ciesin.columbia.edu/docs/001-009/001-009.html*

Worldwatch Database (1998). Available at *http://www.worldwatch.org/*

Worldwatch Database (2000). Available at *http://www.worldwatch.org/*

World's Worst Polluters (2000). Mines and Communities Website at *http://www.minesand communities.org/Action/press1254.htm*

2

Critically overpopulated planet

Artificial agrocenoses as a launching pad for demographic growth—From the Neolithic to the Industrial Revolution—Napoleonic Wars mirrored by demography— Stabilization in developed countries vs. demographic explosion in developing ones— Between K-population and r-population strategies—The "Energy Secret" of humanity's limitless growth—Will there be enough time for completion of the demographic transition?

Homo sapiens and its hominid predecessors required a few million years in order to increase their numbers from a few hundred thousand individuals to 7–10 million (Kapitsa, 1995). In this time our evolutionary ancestor went from simple gathering to gathering complemented by hunting and fishing, for which corresponding technologies had to be developed. These technologies made possible migration from the subtropics, where only the gathering lifestyle had been possible, to regions with more severe climates. This global expansion was also aided by the taming of fire, the ability to make clothing out of animal hides, and the use of shelters against inclement weather.

All of this enabled our prehistoric ancestors to increase the extent of their habitats significantly, settling all over Eurasia and reaching Australia approximately 50,000 years ago. By about 12,000 to 11,000 years ago (at the end of the last ice age) humans crossed the Bering land bridge to North America from Siberia. As a result, virtually the entire planet became man's ecumene. Furthermore, the non-genetic method of information transmission (i.e., the use of experience, technologies, and cultural skills passed on from generation to generation) gave humans an exclusive edge, turning them into monopolists among all living things.

A particularly important role at this early stage was played by the use of throwing weapons: bow and arrow, spear, javelin, etc. Their wide proliferation around 20,000 to 30,000 years ago is the reason this time is known as the Great Drive-Hunting

Period: drive hunting by driving animals off a precipice became the method of choice, and its main objective was large herd herbivores.

However, ecological monopolies can have unpredictable consequences for all concerned, including the monopolizing species themselves. Thus, as the ice began to melt and retreat from the Eurasian and North American landmasses around 15,000 to 12,000 years ago, radical changes in the landscape caused the extinction of the so-called "mammoth fauna". Some researchers believe that this undermined the economic basis of the Neolithic hunters, leading to severe food shortages. Indeed, as paleoanthropological studies indicate, at the end of the last ice age a severe drop in the number of Neolithic encampments is observed in some areas. From this follows a hypothesis that many groups of hunters may have starved to death, and the planet's human population may have dropped by a factor of 10 according to some estimates! However, in all likelihood this was a rather gradual process, and the high degree of flexibility characterizing these early hunters, as well as the considerable cultural experience accumulated by that time, made it possible for human beings to start hunting smaller herbivores. The latter were quick to fill the huge territories freed as the ice sheets retreated in the last melt.

At the same time, in areas characterized by a warmer climate and long-standing gathering traditions, the cultural evolution of man proceeded in a completely different direction. In this connection, efforts were made to become as independent as possible of nature's vagaries. This happened approximately 12,000 to 10,000 years ago (just after the last glaciation); that is, at the very end of the Neolithic when humans began to practice arable agriculture and animal husbandry. This fateful turn of events is known as the Neolithic or Agricultural Revolution.

According to N. I. Vavilov's hypothesis, the geographic primacy of modern agriculture belongs to the river valleys of the Fertile Crescent, namely, the Anatolian Plateau (modern-day Turkey) and the sources of the Tigris and Euphrates rivers. It was there that the culture of wheat cultivation was born as has been shown by the latest archeological findings. Following the end of the Pleistocene and the Drive-Hunting Period, the agricultural technologies that had appeared in the course of the Neolithic Revolution provided the impetus to the second stage of global population growth.

At first, this growth extended only to limited areas of the Eastern Mediterranean, as well as the valleys of the Indus in India and the Yellow River in China, the cradles of the first agricultural civilizations. However, the end result was an almost 20-fold increase of the Earth's human population: from about 7 million to 10 million people 10,000 years ago to 100 million to 200 million by the beginning of the Common Era (Kapitsa, 1995). This growth occurred despite the spread of infectious disease (especially among children) and the associated increases in childhood mortality rates caused by high population numbers, and sometimes by extreme population density in centers of ancient agriculture. However, childhood deaths were more than compensated by rising birthrates facilitated by rising longevity and a corresponding prolongation of women's child-bearing years: beyond 25–27 years.

Under these conditions, agricultural peoples developed the notion of realizing a woman's full reproductive potential, as exemplified by God's famous command from

Genesis 1:28: "Be fruitful and multiply." However, of the typical six to eleven children born per woman, only two to three survived. In this connection, V. Dolnik writes: "As opposed to us, our ancestors had many siblings, but in the graveyard rather than in living proximity" (1992). Therefore, at some point, high birth rates began to lag behind the growth in childhood mortality, which explains the subsequent slowing down of population growth on the planet. From the beginning of the Common Era to 1650 (the approximate time of the Industrial Revolution's origins) the human population grew from 100–200 million to 450 million (i.e., only a three-fold increase).

The third stage in global population growth (which is still going on today) was launched by the Age of Discovery and Exploration. This was coupled to improvements in the dwellings of the Europeans, the spread of hygienic principles, and the gradually rising level of education. The geographic discoveries of the great explorers stimulated the wide proliferation of new agricultural technologies on a planetary scale. Unheard-of agricultural crops were brought to Europe from the Americas, thereby expanding the Old World diet far beyond its usual range.

However, the greatest transformation of Europe was made possible by the Industrial Revolution in the late 18th and early 19th centuries. Soon thereafter science and technology made a giant leap forward. Along with the appearance of liberal economies and civil social structures, fundamentally new technologies (industrial, agricultural, and medical-hygienic) began to facilitate not only increases in the human lifespan but, even more significantly, decreases in child mortality.

The unchanging high birthrate trend led to quick population growth that gradually became truly hyperbolic (Kapitsa, 1995; Cohen, 1995). Thus, while at the beginning of the Common Era the population growth rate did not exceed 0.04% per year, by the end of the Industrial Revolution that rate reached 0.3% (i.e., a 7.5-fold increase!). This created a demographic crisis in Europe, and its partial resolution was made possible by waves of emigration to the New World, Australia, and Siberia.

The associated surplus of labor force created a basis for a rapid economic boom, since pay rates could be kept at or below the subsistence level. Because the population was predominantly young, armies could easily be formed, and conquests were made by a broad generation of "passionate men", to use a term from the work of the Russian ethnologist L. N. Gumiliov. These new men of action strove for quick social change. Turbulent demographic processes were in many ways the basis for a wave of military and revolutionary upheavals within Europe itself, as well as colonial conquests and world-changing struggles on other continents. Europe passed its peak of population growth rate in the 19th century.

By 1900 the Earth's human population reached 1.6 billion, progressing at a rate of 0.5% per year. Of that number, colonies and dependencies had 1,070 million people while the developed world (including the Russian Empire and Japan) was home to 560 million (*World Resources 1990–91*, 1990). At the same time, the 20th century in Europe and North America was characterized by a gradual stabilization of population growth. The new population formula (i.e., low birthrate, low mortality rate, high longevity) was something like a belated reaction to the drop in child mortality as a

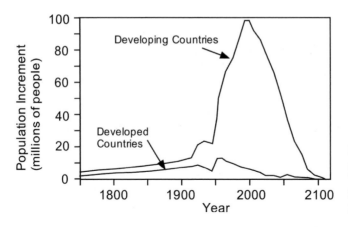

Figure 2.1. The demographic transition. The global population increment from 1750 to 2100.

result of improved living conditions, as well as progress in medicine and hygiene. In the developed countries the gradual decrease in the birthrate (1–2 children per family) has been continuing to this day, which has led to a sharp drop in the growth rate of the native population. In these countries, population growth was continued almost exclusively through immigration.

However, the demographic dynamics in certain parts of the developing world in the 20th century unfolded in a completely different manner. Thanks to progress in agricultural science, the implementation of new medical and hygienic standards, as well as access to clean water, child mortality started to decline while longevity kept growing. In Europe, which was the source of the technologies that allowed humanity to win the first round in the struggle with death, this process had expanded over two to three centuries. In the developing world, however, the same developments required only a few decades, which was essentially the cause of demographic destabilization.

As a result the developing world was considerably late in its shift to the European "high birthrate, low mortality rate, increased longevity" population formula which had already been discarded by Europe. At the same time, this transition was much more explosive, as evidenced by the fact that population growth in the developing countries reached 2.6% per year by 1967, which had never been the case in Europe. In fact, 1967 turned out to be a record year with respect to population growth in the world as a whole: 2.1% per year. And as of 1971, this rate began to decline, reaching 1.6% by the end of the century.

The last decade of the 20th century became a demographic turning point. A demographic transition process began to take place as the "decreasing death rate, birthrate, and population growth" formula, which had already manifested itself in developed countries in the late-19th century, gradually began to prevail on average in the world at large.

Despite the gradual decrease in the birthrate, absolute population growth in the world in the 1990s remained, alas, the highest in all of human history: 80 million–92 million people per year. The explanation is the relatively slow drop in birthrate as compared with the mortality rate. Furthermore, the population of certain Third

World countries (i.e., almost all of Africa and many Moslem societies in Asia) has continued to grow in relative terms as well (i.e., the birthrate in these regions is still increasing). In this manner, the yearly population growth rate in Africa from 1950 to 1990 rose from 2.6% to 3%, in Iran from 2.8% to 3.4%, in Pakistan from 2.6% to 3.6%, and in Saudi Arabia from 3.6% to 4.1%. In population-heavy India, the growth rate had decreased by a mere 0.1% in the 1980s (*World Resources 1990–91*, 1990). The population growth rate remained at 1.6% in 2008. Absolute population growth on Earth reached its peak at the very end of the 20th century, but it will not stop, according to demographic estimates, until the middle of the 21st century at the earliest.

It is obvious that regions suffering from the demographic explosion consist almost entirely of poor, underdeveloped countries. Constantly consuming increments in production, this form of population growth condemns affected areas to "running on the spot", as J. Neru put it, and often leads to a drop in an already squalid standard of living. However, this is the nature of the momentum characterizing demographic growth that

> "is controlled by biological mechanisms, as well as a very complex population system supported by daily customs, traditions and religion. A population needs time—a few generations—in order to adapt the birthrate to a new death rate level" (Dolnik, 1992).

However, the question is whether humanity has this time at its disposal.

The impact of the demographic crisis is not limited to food shortages and economic problems in the Third World. As was once the case in Europe, the structure of the population in the developing world has changed drastically, and over a historically short period of time the proportion of young people has become dominant. This generation produced the leaders of the anti-colonial movement in a part of the world where until the late 1940s, much of the territory was under European control. Political parties were created, some of which espoused extremist ideals. At the same time, lower officer ranks gained power in a whole series of countries where they established semi-military dictatorships. The ambitions of such regimes played a significant part in unleashing internecine and inter-ethnic wars covering huge territories in Asia and Africa.

As a result, in the last third of the 20th century the percentage of the world's expenditure on the military by developing countries increased dramatically, from 6% in 1965 to 20% by the late 1980s. It ought to be added that a significant part of these countries' foreign debt is determined today by arms imports. And as for the roughly 120 armed conflicts that flared up all over the planet between the period after the end of WWII and the early 1990s, the vast majority took place in the world's poorest regions, causing the deaths of over 20 million people. Essentially, since the end of WWII there has not been a single year without bloodshed due to local armed conflict in one region or another.

Along the same lines, the wave of international terrorism unleashed upon the world in recent years should not be viewed in isolation from demographic issues.

There is no doubt that the lack of encouraging life prospects, the sense of having lost out, the perception of having been humiliated constitutes the perfect breeding ground for today's Bin Ladens. "These young men," as we read in the magazine *Geo*, "were produced in an atmosphere of desperate fury which, according to psychological studies, can lead to an increase in the Narcissus Syndrome. The latter may cause one to lose the normal sense of self-preservation" (Kuklik *et al.*, 2002).

Apparently mirroring the dwindling value of a single child's life in a population with a high birthrate, the value of a human individual's life in general depreciates in those places where this excessive birthrate is coupled to overpopulation. According to some researchers (e.g., Severtsov, 1992), our genetic memory retains information regarding the optimal density of our brethren per unit of occupied territory. Repeatedly exceeding this density can have a negative effect on human psychophysical states. In this connection, one cannot rule out factors stemming from an unfavorable demographic situation in impoverished and overpopulated areas of Iraq or Palestine. Here is where we find the distorted psyches of suicide-bombers characterized by a strikingly dismissive attitude toward their own life and the lives of others.

The process of decolonialization has not brought about the expected relief to former colonial peoples, among other reasons because of the explosive demographic leap in the post-colonial period. Decolonialization brought about essentially the same kind of war and revolution phase that Europe had begun to experience two centuries earlier. On the other hand, dramatic overpopulation acted as a spoiler, causing economic drag and acting as a powerful source of stress on the environment with all of the associated ecological and social consequences.

However, let us return to the world's demographic situation. As we enter the third millennium, humanity has passed the 6 billion mark. It had taken more than a million years to reach the first billion. By 1820 a billion people lived on Earth, and in the next 107 years (by 1927) this number doubled. Only 33 years were needed to reach the third billion in 1960. Fourteen years later another billion was added, and then another in 13 years. The last (sixth) billion came 12 years after that, in 1999. Only in the last two decades of the 20th century did this wild growth begin to abate gradually, dropping to 1.3% by the year 2000 (Maksakovsky, 2003).

At this point we can correlate these gargantuan numbers with what is happening in nature. It must be remembered that humanity, despite all of its technological might, is but one among many species inhabiting the Earth. And just like any other species, we are fully subject to the laws of biospheric sustainability that determine the limits of population growth. In a balanced environment every biological species has its own range of normal numbers that is determined by laws of stabilization and whose magnitude depends in particular on the size and mass of the organisms in question. Human beings, who fit within the category of large animals, have exceeded their naturally determined numbers by 4–5 orders of magnitude (Akimova and Khaskin, 1994). Today, the biomass of humanity along with its domestic animals and cultivated vegetation has reached 20% of the biomass of all natural species on dry land while at the start of the 20th century that figure was no higher than 1%–2% (Warmer *et al.*, 1996). How could this happen?

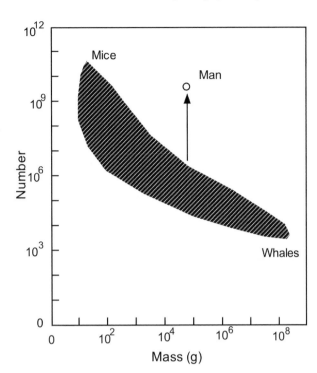

Figure 2.2. Interdependence between heat mass and mammal numbers. (Shaded area) Correlation field between average heat mass of adult individuals from a mammalian species (M, r) and their approximate numbers (N). Excluded are species of raised animals, as well as rare and disappearing species. The arrow indicates the extent to which the modern numbers of *Homo sapiens* have exceeded the "initial" numbers of man's ancestors determined by the law of natural selection. The excess amounts to four orders of magnitude (Akimova, T.A. and V. V. Khaskin, 1994).

Two population strategies, typical of most living organisms, are known in biology. The so-called *r*-strategy characterizes small mammals, for example. Here we find sharp numerical fluctuations in populations that tend to increase and lead to the exhaustion of food resources. This, in turn, causes a drastic drop in the total number of individuals, creating a recurring pattern: population explosion, collapse, stabilization. Repeatedly undermining the very basis of its own existence, a given population appears to keep squeezing through a kind of bottleneck. Examples of such species include the mouse-like rodents (lemmings, squirrels, etc.). This population strategy can be considered as highly entropic since it is associated with a high percentage of organism culling and deaths, as well as the mortification of living substance in a population (mortmass elimination) (Krasilov, 1992).

In contrast to the *r*-strategy, the *K*-strategy typifies first of all the large mammals. It boils down to the maintenance of stable density and numbers in a population by means of low fertility and death rates coupled with a given individual's relatively high longevity, the latter being attributable to low vulnerability (Severtsov, 1992). An example of the *K*-strategy among tundra wolves is provided by the Canadian naturalist Farley Mowat:

"Until they are of breeding age most of the adolescents remain with their parents; but even when they are of age to start a family they are often prevented from doing so by a shortage of homesteads. There is simply not enough hunting territory available to provide the wherewithal for every bitch to raise a litter. Since an overpopulation of wolves above the carrying capacity of the country to maintain would mean a rapid decline in the numbers of prey animals—with consequent starvation for the wolves themselves—they are forced to practice what amounts to birth control through continence. Some adult wolves may have to remain celibate for years before a territory becomes available" (Mowat, 1963, p. 130).

Obviously, the *K*-strategy should apply to humans. However, in some regions and certain historical periods something entirely opposite has been observed, namely, a shift toward the *r*-strategy. One need not look far for an explanation. Simply put, the human habitat has been expanding artificially as people conquered more and more territory by means of mastering new, constantly developing technologies. One of the first such technological discoveries, drive hunting, may have already played a nasty joke on human beings. Intensive hunting or over-hunting sped up the natural process of extinction among large predators and animals from the mammoth fauna. As a result, the primeval hunters undermined their own trophic resource, thereby endangering the survival of the entire population toward the end of the last Ice Age.

However, since the time when humans learned to create agrocenoses, their food supply became a hundred times more reliable. The inexorable growth of this supply has at times kept up with and at times lagged behind population growth. Very recently (in the 1950 to 1960s) a significant role in supplying food to the populations of developed and developing countries was played by the so-called "green revolution". The latter's most decisive participants were the European States, the U.S., and Canada.

Implementation of high-yield crops and the latest irrigation methods, increased efficiency and mechanization of agriculture, while the use of fertilizers and agricultural chemicals all sharply increased the output of agricultural production. Thus, in the second half of the 20th century, it became possible to increase the world production volume of meat and fish five times, soy beans nine times, and grain almost three times. This pushed back the shadow of famine from the lives of millions in the Third World. From 1950 to 1984, the production of grain on Earth grew at a rate of 3% per year, which outpaced population growth. At the same time yearly per-capita grain consumption increased from 247 kg to 342 kg (Brown *et al.*, 1999).

Of course, obtaining these mountains of food would have been inconceivable without a simultaneous increase in energy use. Indeed, there are hardly any areas left where the soil is tilled by oxen, and as a consequence, all agricultural technology requires more and more hectoliters of fuel. The processing, transportation, and storage of agricultural products are also inseparable from energy expenditure. Therefore, in order to increase food production on Earth by only 2% per year, energy use has to climb by no less than 5%.

And yet, this is only the tip of the iceberg; food production accounts for only 10% of humanity's energy needs. Humans are the only living beings on the planet capable of using energy sources other than food. Thus, the lion's share (i.e., 90% of our energy expenditures) is used in satisfying specifically human needs: heating and lighting living space, transportation, the mechanization of labor processes, leisure, etc. (Rabotnov, 2000).

This is why the rate of growth of civilization's consumption of energy has greatly outpaced the current rate of population growth. From 1850 to 1987, a period of 140 years when the Earth's population grew four times, our energy consumption went up 1,000 times! (Shelepin et al., 1997). And this energy (rather than demographic) boom occurred almost entirely in the 20th century.

At this point we should consider the relationship between this river of energy and the sources feeding it. Only 2.7% of the world's energy consumption relies on renewable sources of which 2.4% are hydroelectric stations and only 0.2% are wind, geothermal, solar, and other installations from the so-called alternative energy category. The remaining 97% relies on non-renewable energy sources: oil (44%), gas (26%), coal (25%), and nuclear power (2.4%) (Chow et al.: 2003). And it is in these energy sources unknown to nature that the proverbial "Excalibur" is located, the secret of humanity's power today. This is what allows people to go with deceptive ease beyond strict species-based limitations imposed on all living organisms by the biosphere.

However, today our "limitless" energy growth appears to have reached its limit. This is not only because we have evidently reached a dead end in the decades-old quest for energy from nuclear fusion and continue to cover most of our energy needs with non-renewable (i.e., fundamentally finite) fossil fuels. Further increases in energy use cannot keep pace with population growth, and this goes beyond resource limits since our total energy production has approached a critical threshold beyond which we risk destabilizing the Earth's climate irreversibly (more in Chapter 14). Perhaps even more noteworthy is the fact that energy investments, as in the case of financial investments in agricultural production for that matter, are no longer characterized by the high returns of the past.

About 30–35 years ago (i.e., back in the "pre-ecological period") a popular play was produced on the Soviet stage. Entitled *A Mesozoic Story*, it dealt with people working in the oilfields of Baku in Azerbaijan. The protagonist is a geologist possessed by the idea of extracting oil from the depths of Mesozoic strata. In the hope of obtaining funds for another exploratory blast, he goes to see an old friend who is a powerful boss in the Baku oil trust. When the geologist enters the boss's office, its cautious and relatively sober occupant, who has little faith in the rhetoric of his tiresome petitioner, tries to evoke the unfortunate consequences stemming from the previous blast. The boss refers to the tons of fish that ended up swimming belly up at the water's surface. In response, the geologist uses his most convincing argument: "Fish can multiply whereas oil cannot."

And so, to pick up the words of the play's protagonist with a sad twist, today the metaphoric fish is not multiplying anymore; that is, crop yield growth appears to have reached its saturation point, and the use of additional fertilizers along with more agricultural chemicals no longer keeps driving up the numbers. Since 1985 the growth of world grain production has been declining. Thus, in 1985–1995 that growth was no greater than 1% per year (Dyson, 1995). As for rice production in rice-harvesting

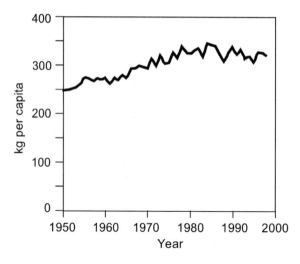

Figure 2.3. World grain production per capita, 1950–1998 (U.S. Department of Agriculture, USDA).

regions, it went up by only 1.5% per year in the 1990s whereas the yearly population growth in these areas was 1.8%.

The key issue here is that despite the impressive success of agricultural technologies in expanding food production, this expansion has, of late, been unable to keep up with the rapid population growth of the developing world. Starting with 1984, regardless of continuing growth in grain production, the per-capita amount of grain produced began to decrease by approximately 1% per year, reaching 319 kg per year by the end of the 20th century (Brown *et al.*, 2000).

In certain countries (e.g., India) this number fell to 200 kg by the end of the century. In the case of India the yearly amount of grain produced per person was lower than a hundred years earlier; that is, in colonial India where it had been 250 kg. This is a very bad symptom indeed.

> "Of the 4.4 billion people in developing countries [...] about a fifth do not have enough dietary energy and protein, and micronutrient deficiencies are even more widespread—with 3.6 billion suffering iron deficiency, 2 billion of whom are anemic" (UNDP Human Development Report, 1998, p. 50).[1]

In the same period, systemic malnutrition afflicted around 1 billion people of whom 20 million lived in India, 180 million in Sub-Saharan Africa, and approximately 170 million in China (Hinrichsen and Robey, 2000).

All of these data testify to a food crisis that until now, has been relative rather than absolute. However, the situation is exacerbated by the uneven nature of agricultural investments; that is, the population of some countries suffers from overeating while whole nations (1/5 of the world) live in hungry misery. Furthermore, it is easy to

---

[1] More recent reports are available at *http://hdr.undp.org/en/reports/global/hdr2007-8/*

identify signs of environmental exhaustion and depletion that affect agricultural productivity among other things.

And yet, given the striking progress of biotechnology, experts at the American National Intelligence Council concluded that the prospects for world grain production were good until the year 2015. Serious food shortages, according to their opinion, are likely only in politically unstable regions where conflicts and natural disasters can impede help for victims of famine (Global Trends 2015, 2000). Let us not quibble with the experts who surely have their reasons for the above-cited optimistic outlook. However, even if humanity does manage (and this would not be the first time) delay the impact of the limits determined by biospheric sustainability laws, let us recall the flipside of the medal. After all, the ability to feed a rapidly growing population that is far from numerical stabilization cannot help but lead to further environmental degradation. The environment would then come even closer to a point of no return.

We know what happens to plant or animal species whose numbers reach a critical level in an ecosystem, thereby threatening the ecological balance: population slowdown or catastrophic collapse will bring the numbers of organisms into line with environmental resources. However, sometimes the entire ecosystem dies.

Such situations have arisen in different forms throughout the history of practically every biological species, from bacteria to large mammals. This includes predators that wipe out entire populations of their prey or hoofed animals that trample all of the consumable vegetation in their habitat, etc. However, even though the population numbers of some species go down under the direct influence of so-called ultimate factors (i.e., hunger, environmental deterioration, epizootic outbreaks; populational r-strategy), other species carry in their genes the ability to stabilize their numbers in advance as a reaction to signals of impending overpopulation. An example of such regulation is the mating instinct of wolves during periods of relative shortages in a given hunting area, as in the case of tundra wolf behavior described by F. Mowat (see bottom of p. 31).

The question is where humans fit in here; that is, is our biological nature receptive to advance warning of such overpopulation signals? First, as V. R. Dolnik points out, such genetic mechanisms act primarily at the level of entire populations, whereas they are virtually invisible in a given individual. Therefore, no matter how much particular people may be aware of the problem, only mass (statistical) manifestations of overpopulation signal perception can be effective (Dolnik, 1992).

The same author draws certain parallels between the way animals react to overpopulation signals and various behavior patterns in some overpopulated human groups. Notably, animals tend to become more aggressive during periods of external disequilibrium and high population density, which could be compared with analogous trends among some people in similar situations. And just as animals tend to neglect their young in excessively dense natural populations, so too under slum conditions (especially in the developing world) we observe rising numbers of abandoned children and the implosion of stable family structures. Finally, we find so-called collapsing agglomerations in which animals lose all interest in fighting for territory and assemble into dense, often wandering groups where the reproductive process comes to a virtual standstill. In this case Dolnik draws a certain parallel with extreme urbanization whereby people, crowded together in gigantic megalopolises, exist in a kind of demographic black hole. Here, the birthrate begins to drop significantly by the second generation. By no means will everyone agree with such an analogy, arguing that over thousands of years of life in society, humans may have lost the corresponding genetic predispositions. However, one fact remains undisputable: demographic growth has definitely begun to slow down.

Professor A. Antonov, chair of the Department of Family Sociology at the University of Moscow, also points out that the process of population growth in Russia, for example, is slowing down. The mechanisms that used to govern high-birthrate norms are being eroded by such factors as late marriages, more rational attitudes toward sexuality, the contraceptive revolution, free access to abortions, premarital relations, and divorce. In certain instances this undermines the "connectedness" characterizing marital and reproductive behavior (Antonov. 2002).

Until now, the transition to the populational K-strategy has occurred in the relatively well-to-do regions of the world or in locations (such as Russia) where a sharp drop in population can have very negative consequences. However, in other regions of the world, the opposite is taking place; that is, a demographic explosion: the Earth's human population will only stop growing in the second half of the 21st century. That is when, according to some estimates, around 10 billion–12 billion people will be living on the planet.

The latter is a very unpleasant prospect because the biosphere cannot withstand another doubling of the human population. Human beings would then surely end up facing the inexorable force behind the ultimate factors of mortality, control over which has been viewed as a key humanistic accomplishment of the Modern Era.

However, as long as there are examples of developed countries which have managed to make a virtually painless transition from the "high mortality–high birthrate" formula to the "low mortality–low birthrate" formula, a glimmer of hope remains. Essentially, this great social innovation for which humanity owes a debt to Western civilization, with its social, economic, scientific, and technological mechanisms of mortality reduction, is as much an inseparable part of globalization as are high technology, modern forms of education, or the Internet.

Professor A. Vishnevsky, another Russian demographer, points out that

"willingly following the West's experience of the struggle against mortality, the developing world has a disadvantage in being unable to be as quick to perceive new mechanisms of social birthrate limitation. Without a doubt, however, this is but a pause in an unstoppable historical process. There is no way back, and the developing world will eventually outgrow its self-destructive phase of rejecting Western demographic norms. In the end these societies will (and some are beginning to do so already) follow the well-trodden Western path" (Vishnevsky, 2000).

On the other hand, blindly following the Western way is still no guarantee of the problem's solution. Incidentally, this was grasped as early as half a century ago by

Mahatma Gandhi, the father of the Indian nation. When asked by journalists if, after gaining independence, his country would reach the same level of well-being as Great Britain, Gandhi replied that on the path to this well-being, the English colonial giant had robbed half the world. So how many planets would have to be robbed so that India can catch up with its former colonizers?

Indeed, before shifting to a new population strategy, the developed countries managed to destroy 9/10 of their ecosystems, thereby turning the Northern Hemisphere into a powerful center of environmental destabilization. And then they had a hand in wiping out natural reservoirs thousands of kilometers away from their own territories, turning the rest of the world into their resource base. Therefore, the totality of the Western path cannot be used as an example to be followed by developing nations. In order to respond to the ecological challenge, the developing world must find its own special path that may resemble some, but not all aspects of the path followed by the developed world. The question is whether they will have enough time to do this.

References

Akimova, T. A.; V. V. Khaskin (1994). *The Foundations of Eco-development* (*Osnovy ekorazvitia*). Moscow: Izdatel'stvo Rossiiskoi ekonomicheskoi akamedii [in Russian].

Antonov, A. (2002). "Narodonaselenie Rossii u opasnoi cherty," *Znamia* **5**, 180–199 [in Russian].

Brown, L.; M. Renner; B. Halweil (1999). *Vital Signs 1999: The Environmental Trends that Are Shaping Our Future*. New York: Worldwatch Institute.

Brown, L.; H. French; L. Starke; L. Brown (eds.) (2000). *State of the World 2000*. New York: W.W. Norton & Co.

Chow J.; R. J. Kopp; P. R. Portney (2003) "Energy resources and global development," *Science* **302**(5650), 1528–1531.

Cohen J. E. (1995). "Population growth and Earth's human carrying capacity," *Science* **269**(5222), 341–346.

Dolnik, V. P. (1992). "Suschestvuiut li biologicheskie mekhanizmy reguliatsii chislennosti liudei?" *Priroda* **6**, 3–16 [in Russian].

Dyson T. (1995) "Be wary of the gloom," *People and Planet* **4**(4), 12–15.

Global Trends 2015 (2000). A dialogue about the future with nongovernment experts. Available at *http://www.dni.gov/nic/NIC_globaltrend2015.html*

Hinrichsen, D.; B. Robey (2000). *Population and the Environment: The Global Challenge*. Population Report No. 15 (Series M). Baltimore: Johns Hopkins University School of Public Health.

Kapitsa, S. P. (1995). "A model of the planet's population growth," *Advancements in the Physical Sciences* **26**(3), 111–128 [in Russian].

Krasilov, V. A. (1992). *Okhrana prirody: printsipy, problemy, prioritety*. Moscow: Institut okhrany prirody I zapoved. dela [in Russian].

Kuklik, K.; X. Luschak; K. Roiter (2002). "Doroga v rai, vymoschennaia telami nevernykh," *Geo* **9**, 138–143 [in Russian].

Maksakovsky, V. P. (2003). *Geograficheskaia karta mira*. Moscow: DROFA [in Russian].

Mowat, F. (1963). *Never Cry Wolf*. New York: Dell.

Rabotnov, N. (2000). "Sorokovka," *Znamia* **7**, 155–174 [in Russian].

Severtsov, A. S. (1992). "Dinamika chislennosti chelovechestva s pozitsii populiatsionnoi ekologii zhivotnykh," *Biulleten' Moskovskogo obschestva ispytatelei prirody. Otdel biologii.* **27**(6), 3–17 [in Russian].

Shelepin, L. A.; B. A. Lisichkin; B. V. Boiev (1997). *Zakat tsivilizatsii ili dvizhenie k noosfere.* Moscow: Its-Garant [in Russian].

Vishnevsky, A. (2000). "Narodonaselenie Rossii u opasnoi cherty," *Znamia* **5** [in Russian].

UNDP Human Development Report (1998). Available at *http://hdr.undp.org/en/media/ hdr_1998_en.pdf*

Warmer, S.; M. Feinstein; R. Coppinger; E. Clemens (1996). "Global population growth and the demise of nature," *Environmental Values* **5**, 285–301.

World Resources 1990–1991 (1990). New York: Basic Books, p. xii.

Ykovlenko, S. I. (1992). "Termoiadernaia elektrostantsia vechnyi dvigatel'?" *Znanie-sila* **1992**, 11–21 [in Russian].

Ykovlenko, S. I. (1994). "Problema kachestva energii," *Voprosy filosofii* **1994**, 95-103 [in Russian].

3

The ecological equivalents of modern man

Timofeiev-Resovsky's test: how many earthlings can fit into the natural biospheric cycle?—What price is paid by the environment for the average city-dweller?—How many people can the planet feed in the final analysis?—The likely cost of the "final" solution to the food problem.

One day, sometime in the late 1960s or early 1970s, a famous biologist named N. V. Timofeiev-Resovsky approached the mathematician N. N. Moiseiev, who was the assistant head of the Computational Center at the Soviet Academy of Sciences. Timoveiev-Resovsky asked Moiseiev to create a computer-based estimate of the number of the planet's inhabitants that could fit into the natural substance cycles of the Earth, given current levels of technological development. The request was intended to make a point with a twist.

Moiseiev, who would later make a name for himself through his work on computer models of the nuclear winter concept, was interested at the time in the possibility of a quantitative description of the biosphere. He was also working on the development of the biosphere and society as systemically linked elements. Timofeiev-Resovsky, for his part, was considering the issue of applying computer modeling methodology to biology and was trying to attract mathematicians to this project. And so, finally, a biologist did join forces with a computer scientist.

Moiseiev later wrote:

"I spent quite a bit of time fiddling around with this question—three to four months. One day he [Timofeiev-Resovsky] called me up and asked if I had any results. I told him that the uncertainty level was very high, which would make my answer imprecise. However, according to my calculations, the answer would range between something like two and eight hundred million people. He burst out laughing and said: 'That's almost right! I got 500 million, and I didn't have to do any calculations for that.'"

It turned out that Timofeiev-Resovsky had had the answer all along, but he wanted to see how a professional mathematician would arrive at a conclusion. Moiseiev goes on

"Indeed, only 10% of the energy used by humans comes from renewable sources, i.e., this is energy that participates in a cycle. The rest is provided by the store-house of former biospheres or stocks of radioactive materials acquired by the Earth at the time of its formation. Therefore, in order to spare the Earth's non-renewable stock, to avoid violating the natural substance cycle and to live in harmony with Mother Nature in the same way as all other living species, humanity must either lower its appetite and find new technological foundations for its existence, or it must agree to a ten-fold reduction of its numbers" (Moiseiev, 1997).

And so, as various researchers have argued, the "legal" ecological population limit is about 500 million people. Humanity can exist for quite a long time within these limits without coming into conflict with nature. However, as we continually dip into "illegal" (and most importantly finite) energy sources, we act like a child who has pried open the proverbial cookie jar in grandma's pantry. Thus, we have exceeded the above norm not even 10 times, as would have been the case when Timofeiev-Resovsky asked his question, but 12 times! This is the very anthropogenic or man-made pressure that keeps crushing the environment more and more palpably.

In order to picture in more concrete quantitative terms the cost of maintaining the average human earthling, we need to consider the ecological equivalents of modern man as conceived by modern science. As has been pointed out, about 50 tons of raw material are produced and transported per person per year in the world today. Some 3.6 kW of energy are used to obtain and process this raw material every year. For the same purpose 800 tons of water are used since most of today's tech-nologies are like life itself, that is, "wet". This generates 48 tons of waste and 2 tons of end-products, the latter constituting essentially postponed waste (Arsky et al., 1997).

In order to characterize these quantities as excessive or reasonable, we can employ the method used by Folke et al. (1997) to evaluate the ecological impact produced by the daily life of the Baltic region's urban population. For this purpose we need a diagram consisting of three concentric circles, the smallest of which contains a human figure.

The human figure stands for the city-dweller, and the smallest circle represents the small territory corresponding to the city-dweller's living space: residence, streets, squares, workplace, stores and eating establishments, administrative offices, cultural institutions, etc. This area, occupying 0.1 ha, is equivalent to the epicenter of environmental disruption where natural ecosystems have been wiped out in their entirety.

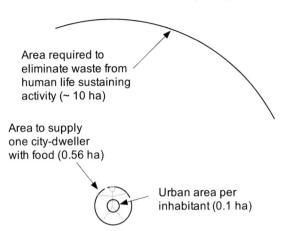

Area required to eliminate waste from human life sustaining activity (~ 10 ha)

Area to supply one city-dweller with food (0.56 ha)

Urban area per inhabitant (0.1 ha)

Figure 3.1. Area required for satisfying the needs of one East European city-dweller. The human figure stands for the city-dweller, and the smallest circle represents the small territory corresponding to the city-dweller's living space: residence, streets, squares, workplace, stores and eating establishments, administrative offices, cultural institutions, etc. This area, occupying 0.1 ha, is equivalent to the epicenter of environmental disruption where natural ecosystems have been wiped out in their entirety. The next circle is larger, representing the area required for supplying our city-dweller with food, natural fibers, and wood. This circle's size depends on the region: for example, in the case of the Baltic Basin—from 0.55 ha per capita (Germany and Scandinavia) to 0.69 ha per capita (countries of the former Eastern Bloc). Finally, the largest circle stands for territory that, without producing anything for our city-dweller, still experiences anthropogenic pressure from the generation of waste products produced by human activity, including biogen emission and CO_2. The area of this territory varies from 4 ha to 10 ha per capita, and its relationship to the area of the disruptive source (essentially urban territory and agricultural land) is roughly 70:1 and 12:1, respectively. This last territory (4 ha–10 ha) is the actual ecological space required to sustain one modern city-dweller.

At this point a small arithmetical calculation is in order: the minimal area required to ensure the life of the above-mentioned average city-dweller (4 ha) multiplied by the number of all city-dwellers on the planet. Since about half the Earth's population lives in urban areas, the result would be 170 million km^2. That is more than the entire dry landmass available on Earth! And this does not take into account the 3 billion people in rural areas, the disruptions caused by industries that serve to increase our comfort level, as well as many other factors. Naturally, the ecological equivalents of man in various countries and regions are quite different from one another, which has to do with their variable levels of economic development and especially their levels of consumption.

In the developed world these numbers are approximately five times above the world average: 250 tons of raw material produced and 16 kW of energy expended per person. In the developing countries the numbers are five times below the world average: 10 tons of raw materials and 0.64 kW of energy per person. And in the poorest of the developing countries the numbers are ten times lower than the world average (e.g., an Ethiopian's ecological equivalent is 50 times below that of someone living in France or the U.S.). This glaring inequality in access to the advantages of life is a major factor that has led to the sociopolitical destabilization the world is presently experiencing. That has been the price for the well-being of economically successful countries.

Let us now reconsider the question posed by N. V. Timofeiev-Resovsky in a slightly reformulated way: What is the maximal number of people that the Earth could feed in the foreseeable future? This is all the more important since the planet's population for the year 2020 may reach as high as 8 billion people.

There are various opinions in this regard, going back all the way to Antoni van Leeuwenhoek (1679). However, most of these views were published in the 20th century, and the estimates vary from 1 billion to 1,000 billion people. The more balanced estimates of recent years range from 2 billion to 20 billion. The question is how to explain the optimism behind the higher numbers.

In 1995, *People and Planet* magazine published a discussion on the topic of the world's ability to feed itself given a population of 8 billion (issue No. 4). Referring to the progress made in grain harvests, technological achievements in the agricultural sector, and the available land potential, many commentators cautioned against excessively pessimistic or alarmist predictions. In other words, the argument was that there were no particular reasons for panic today. What accounts for such confidence?

On the one hand, most such approaches are based on mathematical models where the extrapolation of a population growth curve stems from regional population density estimates, water resource accessibility, the potential productivity of arable land, and other parameters dependent and independent of man. Thus, in a model proposed by J. Cohen from the Laboratory of Populations at the Rockefeller University, the author calculates numerical changes in a population by multiplying the difference between its maximum possible number and the actual number at a given point in time, by a constant called the Malthusian coefficient. Here the maximum population on Earth (the human-carrying capacity) is an uncertain quantity deriving from a series of variable factors. For example, inhabitants that reappear in a certain area can increase accumulated savings or, conversely, "eat up" existing capital; that is, reduce or increase respectively the maximal human-carrying capacity of a given region or the planet as a whole (Cohen, 1995a, b; 1996).

Now, regardless of the mathematical calculations, one should not overlook an ideological component that appears to be present, although we cannot be certain whether this was intentional or unintentional. For example, Cohen refers to a famous statement made by the former American President George Bush Senior, namely, that every human being has hands for working in addition to having a mouth for eating. If we follow this logic, the population limit on Earth appears to be dependent exclusively on human activity, the development of human technologies and, in general, subjective factors of economic growth. One of the participants in the above-cited discussion that took place on the pages of *People and Planet* put it even more bluntly by saying that it was easy to guarantee the food supply. All one had to do was pour money on farmers (Hall, 1995).

Several decades ago another U.S. president, J. F. Kennedy, stood on a high podium and proclaimed that modern humanity possessed the skills and means necessary to eradicate famine from the face of the Earth in one generation. However, Kennedy failed to make this happen in his own lifetime, which was cut short so tragically, nor did those who succeeded him. And so, there are still poor and hungry people on Earth: over 1.5 billion people live on one dollar per day while 35,000 die daily from hunger and poor nutrition (Meadows *et al.*, 1992; World Bank, 1997). Given the 80 million–90 million new mouths to be fed in the world every year, as was the case in the previous decade, all attempts by the international community to feed this hungry army, which is here to stay and which will only keep growing in the next decades, are canceled out.

These depressing numbers are in stark contradiction to Cohen's optimistic position and that of various other authors who believe in the omnipotence of humanity armed with technology that can allegedly dictate its own laws to the biosphere. Admittedly, one can suppose, as the authors of this book do, that the existing land potential, coupled with our agricultural and biological technologies, will continue for a certain amount of time to help feed (more or less acceptably) the current or a larger population on Earth. However, the fundamental obstacle to solving the hunger problem for the planet's billions of people will remain because it resides in the limitations of the Earth's ecological resource.

Returning to the point behind George Bush Senior's above-mentioned statement about eating mouths and working hands, we would argue that the environment is ultimately harmed not by the mouth but by the hands. And this harm is inevitable as we continue to produce food and other goods that provide the basis of human existence. As for the manner in which these "working hands" keep cutting down tropical rainforests or turning the fertile African savannah into desert in order to earn a rather meager living, a mere 9 billion dollars per year for the entire poverty-stricken, famished African continent with all its epidemics and internecine wars, we can turn to a document presented by U.N. experts, namely the 1996 UNDP *Human Development Report*.

All told, the barbaric destruction of the natural environment by the developing countries "earns" them 42 billion dollars per year, which still amounts to far less than the GNP world average of 4,000 dollars per year per capita. In order to reach this elusive GNP target, the economic capacity of these countries would have to increase several times, which would mean a corresponding increase in raw material production, more water and fuel usage, as well as a significant expansion of territories with anthropogenic agrocenoses, etc. Furthermore, in order to ensure a more or less adequate food supply for the almost 8 billion people who, according to various forecasts, may be living on Earth by 2025, the Food and Agriculture Organization of the United Nations estimates that the volume of the world's food production will have to double (UNFPA, 2001). Admittedly, technologically or financially speaking, this problem appears quite solvable. However, even if one day the Earth's hunger problem were to be suddenly solved, then, as Gretchen Daily so aptly puts it, this would be a Pyrrhic victory, achieved at the price of environmental collapse (Daily, 1995).

There continue to be optimistic and pessimistic views on world population. For example, the World Population Balance Organization said:

"Current global population of over 6.6 billion is already two to three times higher than the sustainable level. Several recent studies show that Earth's resources are enough to sustain only about 2 billion people at a European standard of living. An average European consumes far more resources than any of the poorest two billion people in the world. However, Europeans use only about half the resources of Americans, on average ... To become sustainable with Earth's resources, what are our choices? Reducing overall consumption by 25% would do it for now. Or, reducing population by 3 to 4 billion would do it. It's more likely that a combination of both—large declines in consumption and human numbers—will be necessary."

However, the latest report by the U.N. (UNPF, 2007): *Unleashing the Potential of Urban Growth* provides a more upbeat view:

"While the world's urban population grew very rapidly (from 220 million to 2.8 billion) over the 20th century, the next few decades will see an unprecedented scale of urban growth in the developing world ... Cities also embody the environmental damage done by modern civilization; yet experts and policymakers increasingly recognize the potential value of cities to long-term sustainability. If cities create environmental problems, they also contain the solutions. The potential benefits of urbanization far outweigh the disadvantages: The challenge is in learning how to exploit its possibilities."

References

Antonov, A. (2002). "Russia's population at a dangerous threshold," *Znamia* **5**, 180–199 [in Russian].

Arsky, Yu. M.; V. I. Danilov-Danil'yan; M. Ch. Zalikhanov; K. Ya. Kondratyev; V. M. Kotlyakov; K. S. Losev (1997). *Ekologicheskie problemy: chto proiskhodit, kto vinovat i chto delat'?* Moscow: MNEPU [in Russian].

Cohen, J. E. (1995a). "Population growth and Earth's human carrying capacity," *Science* **269**, 341–346.

Cohen, J. E. (1995b). "How many people can the Earth support?" *The Sciences* **6**, 18–23.

Cohen, J. E. (1996). "How many people can the Earth support?" *Population Today* **24**(1), 4–5.

Daily, G. (1995). "Foreclosing the future," *People and Planet* **4**(4), 18–19.

Folke, C.; A. Jansson; J. Larson; R. Constanza (1997). "Ecosystem appropriation by cities," *Ambio* **3**, 167–172.

Hall, D. (1995). "Providing energy and food for all," *People and Planet* **4**(4), 15–17.

Meadows, D. H.; D. L. Meadows; J. Randers (1992). *Beyond the Limits: Confronting Global Collapse, Envisioning a Sustainable Future.* Post Mills, VT: Chelsea Green.

Moiseiev, N. N. (1997). *Kak daleko do zavtrashnego dnia.* Moscow: MNEPU. Chapter 10 "Epopeia iadernoi zimy." Section "Nikolai Vladimirovich Timofeiev-Resovsky" *http:// www.mnepu.ru/go.php?n=54&aa1=109&aa2=2* [in Russian].

UNDP (1996). *Human Development Report.* Available at *http://hdr.undp.org/en/reports/global/ hdr1996/chapters/*

UNFPA (2001). *State of the World Population 2001. Footprints and Milestones: Population and Environmental Change.* New York: UNFPA. Available at *http://www.unfpa.org/swp/2001/ docs/swp2001rus.pdf*

UNPF (2007). *State of the World Population 2007: Unleashing the Potential of Urban Growth.* New York: United Nations Population Fund.

World Bank (1997). *Advancing Sustainable Development: The World Bank and Agenda 21 since the Rio Earth Summit.* Washington, D.C.: World Bank.

WPB (2007/2008). *Current Population is 3× Sustainable Level: Balanced View 2007–08.* Washington, D.C.: World Population Balance. Available at *http://www.worldpopulation-balance. org/BalancedView2007-08/CurrentPopulation3xSustainable*

Part II

Civilization teetering over the abyss of crisis (conclusion)

4

The social dimension of the crisis

Life on $1 per day for 20% of the Third World—Poverty as the foremost pollutant—35,000 hunger fatalities per day—Migrants as the new proletariat of the developed world—Urbanization as a "natural phenomenon"—The flawed environment of the megalopolis

Neither the ecological nor the demographic situation on the planet can be considered in isolation from the social aspects of the crisis. The latter are no less dramatic and illustrate particularly well the extent to which modern civilization has fallen short of attempts to resolve a whole series of so-called basic or eternal issues.

History records many movements for social equality and justice. The greatest of these in recent times was without a doubt Marxism, but neither the Marxists nor their predecessors managed to go beyond social structures in their attempts to reach the "ultimate". At best the founders of this movement paid a modicum of attention to the question of relative overpopulation. However, they viewed this phenomenon not from a global (anthropological) angle, but rather in concrete historical terms (i.e., in connection to the specific situation of the moment). As for the possible interdependence of environmental degradation and chronic social problems, Marxist thinking never went that far.

And yet, many problems, including today's social concerns, were to a large extent already developing long ago when humanity had transgressed its multimillion-year contract with nature. Therefore, today's ecological crisis can be justly viewed as a socioecological one. At the same time, it must be said that the form of this crisis varies from region to region (i.e., issues at the top of the agenda for one country can occupy a lesser place for another).

For example, in the case of the developing world, poverty is a key social problem. A U.N. report from 1997 introduced the Human Poverty Index for the first time as an indicator applying to people who live on $1–$2 a day. Thus, the number of people in the world living on $1 a day is 1.3 billion, and another 1.6 billion people exist on $1–$2 per day (State of the Planet, 2002). In other words, around 40% of the Earth's population still lives below the poverty line.

Of course, this gradation is by no means indicative of every aspect of poverty, which cannot be measured in terms of money alone. Among other issues to be considered are constraints placed on independent life choices, intellectual limitations stemming from the inaccessibility of schooling, as well as the insufficiency of means for maintaining one's health and vital energy. Finally, poverty does not provide the conditions necessary for basic self-respect and the sense of human dignity.

Poverty and environmental degradation are inextricably linked. Half of the world's poor, especially those living in the developing world, work in agriculture. The land they cultivate is of low quality: steep, arid hills and insufficiently fertile soils cleared by the burning of tropical forests, etc. Given the absence of means for maintaining fertility, as well as keeping salinization and erosion at bay, the land in question is quickly depleted.

The other half of the world's poor lives in urban suburbs with highly polluted water and air, among heaps of garbage and vacant lots. Their dwellings usually lack basic amenities, such as electricity or running water. For heat and cooking, wood is cut down from nearby forests and, at times, from urban plantings.

The high level of pollution in Third World cities is also amplified by the inability of the urban poor to sustain the expenses required for environmental cleanup. As a result, a large portion of solid household waste is not taken away. For example, 30% remains within the city limits of Jakarta, 70% in Karachi, and 80% in Dar es Salaam (*Development and Environment*, 1992). And it is in such locations that we find the world's highest level of water and air pollution which makes the industrial centers of the developed world seem almost like mountain resorts. And so, in the Third World, intestinal disease alone (caused by dirty water) brings about the deaths of around 2 million children per year.

In light of the above it is hardly surprising that many developing countries have already destroyed the natural ecosystems in their territories. In Bangladesh there are simply none left. In India only 1% of the country remains untouched, in Pakistan 4%, in Thailand 7%, in China 20%. That latter figure may seem quite considerable, but most of the territory in question is in deserts and on the Tibetan Plateau.

A constant struggle for land and water resources rages in these countries, and in India in particular, this conflict has taken the form of caste, religious, and interethnic clashes. In Bangladesh the confrontations are at the level of political parties. In Pakistan different clans, ethnic groups, and religious movements are at odds. The situation is somewhat better on the African continent; however, here too there are states where natural ecosystems have been destroyed completely. For example, in Rwanda and Burundi we find nothing but agricultural zones and a quickly growing population. Here the struggle for resources has taken the form of furious tribal warfare between the Hutu and Tutsi tribes.

The developing world is afflicted with the barbaric destruction of forests, especially in Amazonia, tropical Africa, and Southeast Asia. This process is progressing especially rapidly in Argentina and Brazil. In the Philippines, 80% of the tropical forests were wiped out in the last three decades of the 20th century (*State of the World*, 1999; Danilov-Danil'yan and Losev, 2000). And this deforestation occurs not only for economic reasons but also for reasons having to do with daily needs, such as cooking and heating. Timber and brushwood are burned in very primitive ovens, which causes serious smoke and soot pollution within dwellings.

However, it is perhaps hunger that should be viewed as the most ominous symptom of today's global crisis, an affliction periodically or chronically affecting huge segments of populations in the developing world. At present there are 800 million people suffering from chronic hunger, of whom 40% are children under five (State of the Planet, 2002; U.N., 2006). Admittedly, hunger has always shadowed humanity, including the Europeans. However, at the threshold of Modernity, Europe was literally saved by the great geographic discoveries that made it possible for excess population to emigrate and take possession of new lands in the open spaces of the Americas, Australia, and Siberia.

Now, however, the stream of migration has turned back, from Africa, Asia and Latin America, into the developed countries of Europe, the U.S.A., and Canada. As in the past, mass migration is driven by hunger and poverty, as well as associated armed conflicts. "For Western Europe, Australia and North America, the growth in migration in the last decade was almost entirely concentrated in flows from poor to rich countries. [...] Today, almost 1 in 10 people living in those countries was born elsewhere" (UNDP, 2004). And nearly 9% of the world's migrants (16 million people) are refugees whose concentrated presence in the areas concerned contributes to political and social tension.

The social situation in the developing world is characterized by other, no less eloquent indicators. For example, around 40% of the world endures without running water or sewage services, and a large proportion of these people occupy city suburbs (Figure 4.1). Hundreds of millions exist essentially without any medical care and have

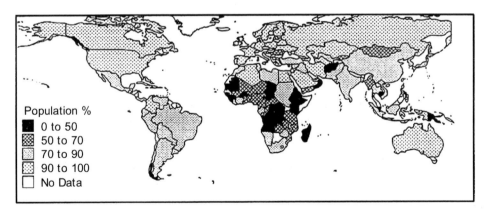

Population %
- ■ 0 to 50
- ▨ 50 to 70
- ▦ 70 to 90
- ▨ 90 to 100
- ☐ No Data

Figure 4.1. Percent of population with access to safe water supply (2000).

no access to schooling. In 51 out of 177 developing countries under scrutiny, the proportion of illiterates among adults (older than 15) ranged from 25%–80% in 2002. And out of 680 million children under 15, 115 million (of whom two-thirds were girls) had no access to education (UNDP, 2003).

The per capita level of GDP (Gross Domestic Product) is used as a universally accepted indicator of social development. The huge GDP gap among different countries and entire continents is one more piece of evidence pointing to the profound malaise of the modern world. In 2002 the per capita GDP of the world's five richest countries was on average 250 times higher than that of the five poorest, and this discrepancy keeps growing. Of the 177 countries under consideration, Luxemburg had the highest GDP ($47,354 per capita), which was 526 times higher than the country with the lowest GDP ($90 per capita)!

It is particularly noteworthy that the countries with a low or very low per capita GDP level include such major Asian States as China, India, Pakistan, and Bangladesh where almost half of the world's population resides. In 2003 the yearly per capita income was as follows: China $989, India $487, Pakistan $408, and Bangladesh $351. The per capita GDP gap between these four countries and the four richest ones was over 70 times (UNDP, 2003). Finally, the personal fortune of a few hundred people at the top of the world's wealth ladder exceeds the sum total income of the countries containing 40% of the Earth's population. In the final analysis, all these contrasts and oppositions feed the ocean of poverty and despair lapping at the shores of the Golden Billion, an island where relative well-being is the norm.

However, it would be naive to assume that the world's social crisis has not affected economically successful states and that a high per capita GDP level is a kind of insurance policy against all social problems. Perhaps the foremost of these problems is the ever-present tension between the local populations of these countries and the so-called new proletariat consisting of migrants who in recent years have been flooding into the U.S.A. and Europe. It is especially in Europe, where so many people have been arriving in search of elusive Lady Luck, that this issue is most palpable.

As Andrei Krivov (2003) writes, in France alone there are 5 million Muslims of whom the great majority are extremely poor and come from North Africa. The first wave of these people who arrived after the end of WWII faces not only open racism, but also a clear unwillingness on the part of the locals to share their place in the sun with these "outsiders". "Only a few representatives of the numerous immigrant communities have managed to work their way up in this society frozen in its caste-like intransigence" (U.N., 2004). However, if the first migrants put up with their meager lot, perceiving it in contrast with their past experience as an enormous advantage of life in Western civilization, their children have turned out to be less patient. Having attended European schools and watched European television, these people have grown accustomed to viewing Europe as their motherland and are by no means ready to repeat their parents' experience. They do not accept being fenced into ghettos allotted to their kind in the French city suburbs where they end up living in grim concrete multistory boxes. Krivov (2003) continues:

"Not having received a good education in their ghetto schools, having no influential or rich relatives, [...] these [young people] are doomed from the start to follow in the footsteps of their parents as cleaners and custodians. Seeing the premature aging of the older generation, their mothers' joints aching from arthritis as a result of 8 hours a day spent in damp catacombs where the famous Parisian mushrooms are grown, the new generation refuses to accept that this life is being foisted on them simply on the basis of their birth in such families. The ensuing mass protests are hardly surprising" (U.N., 2004).

Even if we assume that the above is somewhat of an exaggeration, we cannot ignore the scandalous success achieved by France's extreme right-wing leader Jean-Marie Le Pen. In the first round of the 2002 French presidential election Le Pen obtained 16.86% of the votes, and this in a traditionally left-wing country. Even the recent presidential victory of the tough former minister of the interior Nicolas Sarkozy, who referred to the rioting youths of African origin in the French ghettos as "riff-raff", is also an indirect indication of a yearning for a strongman leader who could shield the average well-to-do French citizen from the unpredictable mass of immigrants.

The latter have taken the place of the indigenous proletariat from the period of classical capitalism, and the middle class views them with as much suspicion now as it did then. This stream of migrants from poor African and Asian countries is expected to grow, thereby contributing to the demographic and economic dynamics of the host countries on the one hand, but on the other hand increasing social tension and perhaps stimulating the rise of nationalism in the developed world (*Global Trends 2015*, 2000).

The problems discussed above are concentrated in the urban environment, which brings us to another related topic. It is perhaps significant that the head suicide bomber from the group that brought about the events of 9/11 spent the last eight and decisive years of his life in Hamburg, the second-biggest city in Germany. Could the experience of existence in a modern megalopolis, where the triumph of scientific and technological progress is paid for with the distortion of normal life, have contributed to the twisted mindset of someone like Mohamed Atta? In this connection we ought to consider a major aspect of the current world crisis: namely, the impact of urban existence on modern civilization.

A few decades ago, megalopolises, defined as agglomerations consisting of smaller cities joining together in densely populated industrial areas (e.g., the Ruhr or Donbass regions), existed only in the developed world. However, the developing countries are catching up and in some cases even overtaking the West in this connection, unfortunately primarily in terms of urban population numbers rather than levels of comfort and well-being. Thus, in a whole set of Near Eastern countries (United Arab Emirates, Saudi Arabia, Kuwait, Lebanon) and in South America (Uruguay, Argentina, Bolivia, Chile, Brazil) the urban population has already reached 75%–92% of the total (U.N., 2004). Until recently this population distribution would have been typical only of the urbanized European states, the U.S.A., and Japan. Furthermore, in the case of Third World urbanization, the process is faster than the rate of industrialization, which means that migration from the countryside into the cities keeps increasing the army of the unemployed and swelling urban slums.

The urbanization process in such places as Mexico City, Buenos Aires, São Paolo, Rio de Janeiro, Calcutta, Mumbai, etc. is highly contradictory. While introducing millions of people to the informational and cultural resources of the modern world, the integration of more and more people into city life exacerbates to the extreme the already acute social problems generated by the phenomenon of so-called "catch-up modernization". Therefore, it would not be an exaggeration to say that in only 20–30 years our planet has become transformed from a "global village" (to use Marshall McLuhan's famous term) into a "global city", making urbanization one of the key issues in the current global crisis.

There are many positions on the role of the megalopolis in the life of modern man. Some people concentrate on the advantages, arguing that high population density and a developed infrastructure make production much cheaper, intensify information flow, and speed up innovation. Others, without denying the negative sides of urbanization, view it as a normal and inevitable part of human development. As Academician N. N. Moiseyev put it,

> "It must be clear to us that most of the Earth's population will continue to live in megalopolises. Unfortunately, this is inevitable! [. . .] The growth of megalopolises is a 'natural phenomenon'. Rather than being someone's invention, urbanization is a consequence of social self-organization. [. . .] We must accept this reality and learn to build megalopolises in such a way as to make them livable. Most importantly, life in a *monstro-city* should become a learnable skill" (Moiseyev, 1998).[1]

However, so far this "skill" has turned out to be hard to learn, and in spite of all the temptations of daily comfort, well-developed medical care, and industrial-scale entertainment, life in a megalopolis often appears to be quite a trial for people, both physically and psychologically.

Human health suffers from the environment of many big cities characterized by high industrial and traffic pollution, as well as very limited means of cleaning up this contamination. The territory of most developed countries corresponds to the so-called centers of environmental destabilization (more on this topic in Chapter 15) which generate two-thirds of the world's waste and consume most of the raw materials produced every year. In light of this, it is easy to see to what extent megalopolises act as epicenters of ecological disruption that cannot be neutralized by even the most advanced ecological technology.

[1] This point of view is by no means shared by everyone. Others correctly point to the revolutionary possibilities provided by modern technology and electronic communication, as well as the increasing role of information in the cost of goods. They stress the advantages of fundamentally different settlement patterns which leave room for regular work at home and beyond the limits of great cities (Danilov-Danil'yan, 2001).

These areas suffer from a concentration of gases and dust particles 5–15 times higher than that prevailing over adjacent territory while the condensation of moisture in the air is 10 times higher. The flow of solar radiation reaching megalopolises is reduced by 15%–20% (30% in the winter months); fog is more frequent, not to mention the all too well-known smog problem.[2] Within city limits, cloudy days are 10% more prevalent than elsewhere (Eurostat, 1995).

With respect to human health, the *aggressive* nature of urban contamination is characterized first of all by the action of chemical pollutants that are thought to influence 25%–50% of all disease in many industrial centers. This has to do with the myriad tons of lead, zinc, copper, chromium, and other metals in various emissions, discharges, and forms of solid waste typical of large industrial centers. These contaminants accumulate in the soil and leak into underground water where they form a kind of "geochemical domain". Lead is particularly dangerous among the heavy metals. Apart from acting upon the endocrine and immune systems, lead slows down the physical and mental aspects of child development. A whole array of aromatic hydrocarbons are thought to possess carcinogenic and mutagenic qualities while the acid-forming compounds of nitrogen and sulfur bring about respiratory and broncho-pulmonary illnesses, including bronchial asthma (Krasilov, 1992).

Unfortunately, the aggressiveness of the urban environment is by no means limited to chemical agents. Harm to human health from physical contamination is considerably less well publicized and in many cases less well studied. For example, in U.S. cities alone, noise from vehicles causes $9 billion worth of damage per year. To this we can also add vibration from rail transport, construction technology, and some industrial plants. Furthermore, there is concern about electromagnetic fields (so-called "electro-smog"), ionizing radiation, and other potential health hazards of a physical nature to which the modern city dweller is exposed like a guinea pig in a giant experimental lab.

Electromagnetic radiation is a devious problem because, like radiation in general, it cannot be perceived by our senses, and its negative effects on human health may appear only after considerable delay. It is only relatively recently that a possible connection has been proposed between energy anomalies around high-voltage lines and child cancer. Having examined the dwellings of children who had died of leukemia, American medical researchers have postulated the possibility that the likelihood of such illnesses may be 2–3 times higher for those residing in close proximity to high-voltage lines.

Furthermore, the high concentration of population in limited areas has increased our vulnerability to natural disasters (e.g., New Orleans Flood of 2005) and epidemics (e.g., SARS outbreak of 2002–2003), the latter being facilitated by the increasing accessibility of global transportation. The same can be said about industrial accidents of which perhaps the most notable was the one that occurred in 1984 in the Indian city of Bhopal (population: approximately 1.5 million). Here

---

[2] In 1952 a smog flare-up in London killed 4,000 people and 20,000 suffered from the effects of the toxicants in the air, which is the greatest ecological smog disaster on record (*The World Environment*, 1992).

thousands died (and many others were contaminated) when methyl isocyanate gas escaped from a Union Carbide chemical plant operated by the Dow Chemical Company. We can also mention the numbers of people killed by earthquakes in large urban centers, especially in the developing world, where a high population density is coupled with large and often poorly constructed buildings to create ideal conditions for maximum casualties. This is exemplified by the Tangshan earthquake of 1976 when at least 35% of the population in this Chinese industrial center of one million perished under the rubble of collapsing structures (Bilham, 2004).

Thousands of years ago our distant ancestor looked up at the stars for the first time and must have felt awe before the grandiose spectacle of the Universe. Since then the starry sky has held our fascination captive in its cold and silent grasp. To quote Immanuel Kant, the towering German philosopher of the Enlightenment, "two things fill the mind with ever new and increasing admiration and awe, the more often and steadily we reflect upon them: the starry heavens above me and the moral law within me" (Kant [1992]). And yet, it has been a whole century since the average big city dweller was able to clearly see the Milky Way or anything like a true picture of the night sky because city lights drown out the light of the Universe.

However, this is more than just an esthetic or metaphysical issue. In practical terms, we are talking about harm to human health by pollution from artificial light sources and the associated disruptions of biological sleep and wakefulness rhythms. This has psychophysical consequences (e.g., nightshift work, glaring advertisements and store windows, the dazzling lights of late-night dance clubs and concert events), all this displaces the night and contributes in some cases to insomnia, daytime drowsiness, as well as such common conditions of our time as depression and chronic fatigue syndrome etc. Having "conquered the night", we may have opened yet another Pandora's Box.

Another type of contamination that extends beyond chemical or physical agents is information pollution. Admittedly, a great deal of benefit has been brought to human civilization by easy access to information: dissemination of knowledge, global communication, increased openness in society, etc. However, it is possible to have "too much of a good thing" (i.e., modern means of information transmission), from cell phones to almost limitless TV channel access to the Internet to video games. Not only do they increase the visual, electromagnetic, and sound background—they also overwhelm us with data. This never-ending stream is six orders of magnitude beyond the physiological capacities of human perception and assimilation (Arsky *et al.*, 1997; Danilov-Danil'yan and Losev, 2000).[3] Although, strictly speaking, information overload is not a problem limited to life in big cities, it is a city-generated phenomenon that has transformed the whole globe into one big megalopolis that is oversaturated with data, useful or not.

The flood of electronic information can create an artificial world unto itself which largely displaces reality, as is illustrated in the 1999 film entitled *The Matrix*. The

---

[3] One can wonder how productive Leo Tolstoy would have been in today's "noisy" world given that he found even the presence of songbirds in the house to be a disruption when it came to creative work.

negative aspects of the virtual world apply mainly to children and teenagers whose perception of the real world may be dulled or distorted through constant exposure to the pixel universe.[4] These days some doctors are talking about computer and internet addiction, which is especially unsettling with respect to the prominence of violence content in the gaming industry. Having experienced countless bloody virtual battles, some people may take on a "superman image", whereby the line separating the worlds on either side of the screen becomes blurred. This can in turn affect behavior and the psyche (Goon, 2003). Perhaps the most notorious example of this phenomenon is the Columbine massacre of 1999 in Colorado. The boys who killed 12 students and a teacher in their own high school were avid fans of violent video games.

The problem of reality loss also concerns television commercials and other forms of advertisement where bombardment with sexually based images (used for selling everything from cars to toothbrushes) and a message of immediate self-gratification create a mentality of superficiality and unbridled consumerism. To quote V. L. Levi (2002), "advertisement is a form of psychological assault which acts upon the sub-conscious like a narcotic, inculcating a lust for pleasure at any cost" (p. 375).[5] All of this could have serious implications for humanity's ecological awareness and eco-logical values. To what extent can a person pumped full of glitzy, short-attention-span data be concerned with environmental issues? Why would someone encouraged to buy, consume, and discard worry about the piles of non-biodegradable waste piling up on the outskirts of our cities?

Finally, urban life is often a source of general psychological stress, a topic we touched upon in Chapter 1. Living "in each other's hair" can take its toll. As Severtsov (1992) supposes, information about optimal population density may be stored in our genes, going back to a time when humans lived in small groups as hunter-gatherers. This would imply that the residents of megalopolises can experience a sense of discomfort (perhaps at the unconscious level) just from constantly existing in close quarters alongside countless and often faceless others. Never being alone from the kindergarten to the cemetery, and in many large cities cemetery "over-population" is a serious issue as well, we suffer from what is termed in psychology "the density or group effect". Despite the amazing adaptability and flexibility of the human psyche, some people crack up, which may result in increased family violence, criminality, alcoholism, and drug abuse.

---

[4] There was a New Yorker cartoon showing a father changing a flat tire on a jacked-up car in the rain saying to three children in the car "I can't change the channel; this is really happening!"
[5] This "must have it now" culture is by no means a recent phenomenon, as is demonstrated by the following passage from V. Nabokov's *Lolita*: "She believed, with a kind of celestial trust, any advertisement or advice that appeared in *Movie Love* or *Screen Land* . . . If a roadside sign said: Visit Our Gift Shop—*we had* to buy its Indian curios, dolls, copper jewelry, cactus candy. The words 'novelties and souvenirs' simply entranced her by their trochaic lilt. If some café sign proclaimed Ice Cold Drinks, she was automatically stirred, although all drinks everywhere were ice cold. She it was to whom ads were dedicated: the ideal consumer, the subject and object of every foul poster" (p. 136).

The angst and turmoil of urban life has been described by numerous analysts of the human mind, as well as novelists of the past two centuries. From the psychologist Carl G. Lange to the psychiatrist Sigmund Freud, from the nightmarish world of Franz Kafka to the no-less abject condition of Liudmila Petrushevskaia's characters, from J. D. Salinger to Yuri Trifonov, time and again we see how much of a challenge it is for the "little person" to retain his or her individuality and to avoid being crushed by the *monstro-city* we have created. And so in a number of respects, there is a heavy price to pay for the comforts offered by science and sociopolitical progress. The artificial urban bubble appears to exist at the expense of the natural environment, the human body, and in some respects the human soul.

References

Arsky, Yu. M.; V. I. Danilov-Danil'yan; M. Ch. Zalikhanov; K. Ya. Kondratyev; V. M. Kotlyakov; K. S. Losev (1997). *Ekologicheskie problemy: chto proiskhodit, kto vinovat i chto delat'?* Moscow: MNEPU [in Russian].

Bilham, R. (2004). "Urban earthquake fatalities: A safer world or worse to come?" *Seism. Res. Lett.* (CIRES and University of Colorado, Boulder). Available at *http://cires.colorado. edu/~bilham/UrbanFatalitiesSRL.pdf*

Danilov-Danil'yan, V. I. (2001). *Begstvok rynku: desiat let spustia.* Moscow: MNEPU [in Russian].

Danilov-Danil'yan, V. I.; K. S. Losev (2000). *Ekologicheskii vyzov I ustoichivoie razvitie.* Moscow: Progress-traditsia [in Russian].

Development and Environment (1992). New York: Oxford University Press.

Eurostat (1995). *Europe's Environment: Statistical Compendium for the Dobris Assessment.* Luxemburg: Eurostat.

Global Trends 2015 (2000). A dialogue about the future with nongovernment experts. Available at *http://www.dni.gov/nic/NIC_globaltrend2015.html*

Goon, G. E. (2003). *Kompiuter: kak sokhranit' zdorovie.* St. Petersburg: Izdatel'skii dom "Neva"/Moscow: OLMA-PRESS Ekslibris [in Russian].

Kant, I. (1992). "Introduction: The starry heavens and the moral law," in P. Guyer (ed. and transl.), *The Cambridge Companion to Kant.* Cambridge, U.K.: Cambridge University Press, p. 1.

Krasilov, V. A. (1992). *Okhrana prirody: printsipy, problemy, prioritety.* Moscow: Institut okhrany prirody i zapovednicheskogo dela [in Russian].

Krivov, A. (2003). "Irakskii gambit 'levikh', ili opravdanie Apokalipsisa." Available at *http://www.ir.spb.ru/content2003-II.htm*

Levi, V. L. (2002). *Kak vospityvat' roditelej, ili Novyi nestandartnyi rebenok.* Moscow: Toroboan [in Russian].

Moiseiev, N. N. (1998). *Sud'ba tsivilizatsii. Put- Razuma.* Moscow: Izdatel'stvo MNEPU [in Russian].

Nabokov, V. (1980). *Lolita.* New York: Berkley Books.

Severtsov, A. S. (1992). "Dinamika chislennosti chelovechestva s pozitsii populiatsionnoi ekologii zhivotnykh," *Biulleten' Moskovskogo obschestva ispytatelei prirody. Otdel biologii* **27**(6), 3–17 [in Russian].

State of the Planet (2002). *The Ecologist* **5**.

State of the World (1999). Available at *http://www.worldwatch.org/node/1038*

UNDP (2003) *Human Development Report*. New York: United Nations Development
    Program.
UNDP (2004) *Human Development Report*. New York: United Nations Development
    Program.
U.N. (2004). *World Urbanization Prospects: The 2003 Revision*. Available at *http://www.un.org/
    esa/population/publications/wup2003/WUP2003Report.pdf*
U.N. (2006) *Hunger and Learning*, World Hunger Series. New York: United Nations.
The World Environment 1972–1992 (1992). London: Chapman & Hall.

5

The role of centralized and market economies

The vanishing act of the centralized economy—The ecological record of socialism—Irrational market forces—Trade and finance wars: the conquerors and the conquered—The trillion-dollar debts of the Third World—Competition in nature and in society—The need for market limitations—Externally imposed limitations on the market—Nature's capital as the inevitable growth limiter—Is there a global economic crisis?

Throughout its entire history, humanity has experimented essentially with only two economic systems: the free market and the centralized economy. The former can look back to more than one millennium of experience with various modifications while the latter amounted to a full-scale trial only in the 20th century (although some of its elements had existed already in antiquity). However, with respect to environmental destruction, both systems have yielded deplorable results in the form of the current global ecological crisis.

At first glance it would seem that by controlling all the wealth of a nation and strictly regulating the parameters of social and economic development, the centralized economic system should have been able to provide all the conditions required for rational land use and ecological conservation. However, alas, this did not happen. Advocating the familiar ideology of unlimited growth, the Soviet leadership aimed to overtake the most advanced countries of the world as quickly as possible on the road to industrialization and urbanization. This was done through the merciless depletion of local natural resources and the disregard of elementary human needs. It soon became apparent that in its negative manifestations the centralized system had much in common with the market system, constituting a kind of warped version of the latter. This, however, was characterized by specifically socialist features.

First of all, in contrast to countries with parliamentary democracies, the totalitarian socialist state could ignore local and regional interests in order to pursue its grandiose plans according to a central will. Public opinion was not an issue while the absence of openness (*glasnost*) and freedom of speech made it possible to conceal the real costs behind the "construction sites of the century" and plans for the "transformation of nature".

Thus, the construction of the first hydroelectric stations of the upper Volga energy cascades in the 1930s, along with the simultaneous creation of the Moscow and Rybinsky water reservoirs, led to the partial or complete flooding of several cities, numerous churches, monasteries, and estates belonging to former landed gentry. Seven hundred towns and villages with extremely rich pasture grounds and floodplain meadows also ended up under water. Altogether 150,000 people were displaced from the flooded areas, of whom it turns out 294 perished after refusing to leave their ancestral grave sites (*Russkaia Atlantida*, n.d.). All this took place in secret and was shielded from the public. Similarly, for a long time two nuclear explosions along the path of river flow displacement in the North European part of the U.S.S.R. remained under wraps. These explosions had been carried out in advance, before the approval of the proposed project, on the firm assumption that the undertaking would indeed be approved (Losev *et al.*, 1993).

On the other hand, the extensive nature of the centralized economic model and its low efficiency led to the large-scale waste of natural resources in the ground and elsewhere. These resources were officially state-owned but in fact belonged to no one and ended up frittered away by various groups and government departments. The ecological disaster of the Aral Sea region, as well as the destruction of fragile northern ecosystems in the process of geological prospecting for oil and gas recovery in Western Siberia, constitutes an eloquent example of what happens when no one really owns anything.

Certain historical similarities can be observed between the behavior of the centralized economic system and that of the free market. First, in the interest of accumulating so-called startup capital, both systems mercilessly robbed and exploited huge groups of people and even entire nations. In the 18th century, England squeezed dry its overseas colonies, and peasant lands, and even monastery properties were expropriated. In the United States, the wealth of the South was based on black slave labor in the 18th and 19th centuries.

On the Soviet side, we had the labor camps with their "free" work force, as well as the subsistence level salaries of the average Soviet citizen in the 20th century. This was accompanied by the exploitation of nature, but in the Soviet case the scale and intensity were even greater than in the Western world. Concentrated in the hands of the party leadership, financial resources could be spent on any megalomaniacal project without any accountability. This made it possible to achieve stupendous rates of development in certain sectors. Gigantic dams created with shocking speed by specialized industrial centers, enormous nuclear icebreakers and other larger-than-life undertakings were the pride and joy of the Soviet leadership. To quote the Russian poet A. Tvardovsky's wry words,

> "All we lack is a canal
> visible from Mars …" (p. 137)

The progress of the Soviet defense industry, culminating in the creation of a nuclear shield, was in one way or another spurred on by tough competition with Germany at first and then America. However, no such stimulation was there for other economic spheres. Without private enterprise to motivate the desire for success, the rate of innovation and the quality of work in the rest of the Soviet economy kept declining with time. Having eliminated all room for personal initiative and competition among producers of goods, the socialist system unwittingly signed its own death sentence. The execution was merely a matter of time, and it came with Gorbachev's famous *perestroika* period. The whole world watched in amazement as the seemingly invincible fortress of the Soviet empire collapsed like a house of cards. The people were left with nothing but debts, the biggest of which became the ecological situation in the long term.

It turned out that in the territories of many former Soviet republics (now independent states) the natural ecosystems had been destroyed almost entirely. Most of today's Ukraine, Belarus, Moldova, the Baltic states, and to a certain extent the trans-Caucasian republics (Georgia, Armenia, and Azerbaijan) are among countries with missing or barely preserved natural ecosystems. In many ways the same picture is observable in the Central Asian States (Kazakhstan, Uzbekistan, Tajikistan, Turkmenistan, and Kyrgyzstan) where only small undisturbed areas remain. However, here the situation is made even more acute by the ecological disaster of the Aral region, affecting primarily Kazakhstan and Uzbekistan. This problem has affected more or less all of Central Asia, profoundly destabilizing the ecological balance of the area. The complete elimination of the consequences (if that is at all possible) will require huge expenditures and an amount of time spanning several generations.

The Russian Federation has been somewhat luckier in this respect. This is due to the particularly intensive drive by the centralized system to modernize the peripheries of the Soviet empire. Since modernization equaled ecological destruction, and since there were fortunately not enough resources for "modernizing" the center of the empire to the same extent, more elements of the Russian ecosystem were spared (by default). But even here, the natural ecosystems in 35% of Russian territory were disrupted (they were almost completely destroyed in 15%). In territorial terms this would equal an area greater than half of Europe, and it ought to be kept in mind that most of Russia's population lives in the territory corresponding to the above-mentioned 35% (Matishov and Denisov, 1999; Losev, 2001).

There have been various attempts to explain the collapse of the centralized economic model. Some have argued that the arms race exhausted the Soviet system. Others point to the crisis of values in Soviet society fatigued by the fruitless pursuit of the communist mirage. However, there is also the notion that the failure of "real socialism" was more the result of a general crisis afflicting human civilization as a whole than the true triumph of the market system. In a crisis situation the latter turned out to be more flexible and therefore more viable, up to a certain point in time (Blanko, 1995).

Whatever one's position, by the end of the 20th century the centrally governed economies were wiped off the face of the Earth astonishingly quickly for the scale of the events in question. Time will tell if this is a temporary or permanent defeat. At the present moment, if we don't count Cuba and North Korea, humanity has only one economic system without any alternatives: the market. And it is the market that will have to bear the whole weight of the problems created by the ecological challenge. The question is whether it is up to this monumental task.

It is universally recognized that the market is blind when it comes to solving long-term strategic problems. Its natural element is the short term or the medium term at best. The market is good at adjusting to changes in profit trends, reacting to current demand and estimating demand levels in the near future. In this context, market success indicators do not necessarily signal positive changes in the quality of life (health, safety, the state of the environment, etc.), and often act as a warped mirror offering a false picture of reality. Thus, half a century ago the U.N. developed the National Count system and its derivative indicator, the Gross Domestic Product (GDP), which frequently establish increases in the volume of a given economic activity under scrutiny. This is also applied to environmental pollution cases, resource depletion, and even accidents (Figure 5.1).

And indeed, medical assistance expenses for a road accident victim, that person's funeral, insurance claims by the relatives, the purchase of a new car: these are all measurable factors of economic activity in the service sector which, as paradoxical as it may sound, contributes to increasing the GDP. And conversely, breastfeeding a child does nothing for GDP while feeding a baby with formula out of a bottle makes a certain contribution to the stimulation of the economy. Similarly, a coronary bypass operation is money from heaven pouring into the coffers of U.S. medical and insurance companies.[1]

One tends to think of the market-based society as the embodiment of democracy and freedom. However, this is true only to a point since even in the most advanced democratic states, private companies and corporations are usually based on a rigidly authoritarian management model. The personnel do not elect their leadership and take no part in vitally important, strategic decisions that may have a serious impact on the environment and similar phenomena unquantifiable in terms of net profit. And so the choice between "the bottom line" on the one hand, and ecological considerations on the other, rests in the hands of single individuals or small, powerful, often-interlocking boards of directors:

" 'There is only one leader.' Philip H. Knight, Chairman and CEO, Nike"

[1] In the 1990s Clifford Cobb and the Stockholm Environment Institute proposed the Index of Sustainable Economic Welfare (ISEW), also known as the Genuine Progress Indicator, which includes not only economic but also quality-of-life components (Weitzsäcker *et al.*, 1997). And indeed, according to estimated data, as of the late 1960s, despite the growth in GDP, the Index of Sustainable Economic Welfare has been moving down. While in 1950 the ISEW was around 70% of the GDP, in 1992 that number had already dropped to 25%. The absolute value of the ISEW in the U.S. has decreased by 30% since 1950. This is an indirect illustration of the extent to which the interests of people, society, and modern business can diverge.

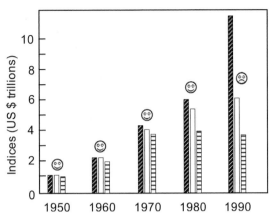

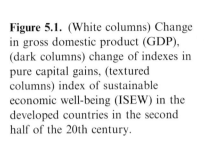

Figure 5.1. (White columns) Change in gross domestic product (GDP), (dark columns) change of indexes in pure capital gains, (textured columns) index of sustainable economic well-being (ISEW) in the developed countries in the second half of the 20th century.

is the succinct epigraph from a book tellingly entitled *The Discipline of Market Leaders: Choose Your Customers, Narrow Your Focus, Dominate the Market* (Treachy and Wiersema, 1995), which leaves no doubt as to who is the boss in most elements of the free market.

Similarly, when it comes to consumption, the freedom of choice turns out to be less "free" than one may assume. To quote A. B. Veber (1999), "Corporations foist desires on people far beyond real needs. The market is a mechanism that creates demand for the unnecessary." The accuracy of this assessment is demonstrated by a truly striking figure from the American National Academy of Engineering: 80% of mass-produced goods are discarded after one use while much of the rest is used no more than twice (Weitzsäcker *et al.*, 1997).

Some companies are openly interested in making sure that their products cannot be used for too long by the consumer (i.e., there is at times an uncanny relationship between the length of the warranty period and the point at which a given gadget tends to break down). Repairs for a number of consumer items are not built into the equation because they are too costly. The proof can be found at city landfill areas brimming with virtually new and still usable equipment that would be repaired elsewhere in the world or under different circumstances. All this is part of "legitimate" strategies aimed at boosting consumer demand. Could the father of the liberal economic system Adam Smith, as he contemplated the harmonious situation created by his famous "invisible hand", have envisaged the struggle for the wallets and "souls" of the consumer taking place in today's market place? And what would he have said about the profit scale driving the shadier sides of the market system, such as the pornography business or the drug trade?

To be fair, the vibrancy of the modern economic system cannot be denied. After WWII the developed world in particular enjoyed unprecedented growth rates. And even in the developing world certain positive changes have taken place. By the 1990s poverty and hunger receded considerably in such major states as China and India. The proportion of children receiving grade school education has grown while child mortality has declined in certain parts of the Third World. This has been accompanied by localized increases in human longevity. However, this is a patchy picture, and in many respects the above-cited developments have been a case of "one step forward, two steps back."

Between 1985 and 1990 the proportion of the poor in the global population as a whole stopped decreasing; after 1990, and especially after the financial crisis of 1997–1998, it began to grow rapidly. The gap between the Third World and the Golden Billion has been growing wider, not the least because of the protectionist policies implemented by the developed world. This is perhaps the most nefarious aspect of the "free" market system today since the rules of global economic engagement are dictated by the rich countries. Relying on subsidies (in the farming sector especially), tariff barriers, licensing policies and quotas, economic giants, such as the United States, have managed to make numerous developing countries less "developing".

The extent to which this is a win–win situation for the haves of this world can be demonstrated by the fact that the developed world receives over three-quarters of the global trade volume. This is where most of the developing world's export activities are aimed, but in a very uneven way since raw materials prevail in this trade. On the other hand, the export of the Third World's industrial production is only 12%–13% of the global volume, and another 7%–8% comes from countries with transitional economies. The other 80% of the global industrial commercial exchange end up enriching the banks and corporations of the developed world. The result is that the per capita GDP indicators are on average at present 10–15 times greater in rich countries than in poor ones (and 50–100 times greater than in the poorest parts of the world).

At the same time the foreign debt of the developing world was over $1,200 billion by the early 1990s. By the end of the same decade the rich creditor states agreed to forgive some of these debts, partially and even fully in the case of the poorest countries. This was accompanied by corresponding conditions, and the process is continuing today. As for the "unforgiven" parts of Third World debt, the growth of interest payments has led to a yearly capital drain calculated in the tens of billions of dollars. It ought to be obvious who the losers are in this scenario that could be qualified as a hidden economic war between the developed and developing worlds.

However, the inhabitants of the developed world are not necessarily always the winners. By endorsing free trade with China, the U.S. allowed its manufacturing infrastructure to move to China so that China's economy is booming while the U.S. economy is perched on the edge of a precipitous drop. Furthermore, wealth is at times more a curse than a blessing. Thus, the passion for gain can be accompanied by addictive consumption beyond the consumer's means. Shopaholism is a serious problem in today's society, turning millions of people into something like gamblers unable to tear themselves away from supermarkets and department stores, not to mention on-line purchasing channels. According to data from the Frankfurt magazine *Neue Zeiten* (2007, No. 79), in the U.S.A. alone there are 15 million shopaholics while 2%–10% of Europe's adults suffer from this condition. This is but one way in which the free market facilitates the mindless waste of resources. Others, which are less easily quantifiable in monetary or economic terms, include the main subject-matter of this book (i.e., environmental pollution, the excesses of urbanization, and technologically mediated accidents).

The list of the market's imperfections, irrationality, and destructive elements can be continued. However, it would be an exaggeration to maintain that the free market system is a hindrance to normal life in every respect and at all times. Even if we turn to the competition among living organisms in nature, which guarantees the preservation, stability, and diversity of species over millions of years, it should be obvious that there is something "natural" about the market as well. The aforementioned "invisible hand" postulated by Adam Smith to describe the balance between supply and demand may be this universal mechanism of life in general, governing the behavior of sticklebacks and corporate CEOs alike.

The free market is based on the activity of living people pursuing their personal interests, as well as the interests of the social groups that they represent. As these players interact on the economic stage, the efficiency characterizing the involuntary reciprocity of their competition is truly striking. No mathematical model nor computer brain has been able to ensure the uncanny precision of the price parity that the market maintains with a virtually "ecological" smoothness. In Adam Smith's immortal words from *The Wealth of Nations*, "it is not from the benevolence of the butcher, the brewer, or the baker that we expect our dinner, but from their regard to their own interest. We address ourselves, not to their humanity, but to their self-love, and never talk to them of our own necessities, but of their advantages" (Smith, 1976). Thus, without trying to do so, Smith's butcher, baker, and brewer end up benefitting each other and society in general as they pursue exclusively personal gain. And the disregard of this feature inherent in human nature was the mistake that led to the inevitable collapse of the centralized economic system. The great conundrum in economics is that a society organized from the top down supposedly for the benefit of the people, does not seem to work as well as one that depends upon the personal ambitions of individuals. Nevertheless, some degree of governmental regulation is needed. The degree to which markets should be regulated is not universally agreed upon. Beginning with the presidency of Ronald Reagan in the U.S., there has been a strong trend in the U.S. (and elsewhere) toward deregulation of utilities, banks, and businesses in general. Unfortunately, the Reagan administration interpreted "deregulation" as "no regulation", leading to the infamous U.S. Savings and Loan scandal of the 1980s, widespread corporate malfeasance in the 1990s (e.g., Enron), and the current sub-prime real estate fiasco.

And so there is no need to reinvent the wheel in search of a solution to today's ecological dead-end: a brand new economic model is not required, given the inability of humanity to come up with a worthy alternative to the market in the thousands of years since the dawn of civilization. However, eliminating some of the market's negative features is a different matter altogether. As was observed by Herman Daly, professor at the School of Public Policy at the University of Maryland and Senior Economist at the Environment Department of the World Bank (1988–1994), when it comes to allocating resources and distributing earnings, the market can be relied upon within ecological and ethical limitations prescribed by instances outside the market itself. Inside the framework of these externally imposed limitations, the market should be free (Daly, 1977). Here are some examples of such limitations outside of which one cannot talk about any sort of sustainability in the economic sphere. The rate of forest clearing cannot exceed the rate of natural forest growth and regeneration. Fishing must not undermine the reproductive capacity of fish stocks. Water use should not disrupt the mechanisms of water resource recovery. Anthropogenic CO_2 emissions cannot exceed the capacity of ecosystems and the World Ocean to absorb this gas, etc. In general, the exploitation of natural systems must not reach levels where the environment can no longer regenerate itself and sustain ecological equilibrium. The question is to what extent such limitations are realistic within today's economic practices. The market system in its current state can only keep reproducing itself with its inherent entanglement of contradictions: extremes of poverty and wealth, hyper-consumption, resource waste, ubiquitous pollution, and never-ending crowding of nature.

In the meantime the ecological excesses of civilization are having more and more economic consequences. The slowing down of something as "natural" as economic growth is no longer taking place exclusively on the basis of financial or production parameters, but as a function of the state in which the environment finds itself. As Herman Daly points out, man-made capital is used to determine the rate of growth; however, today, after the unprecedented increase of this capital, the "capital of nature" is becoming more and more of a limiting factor. This shift in economic focus means that it is not the number of fishing vessels, but the reproductive capacity of fish populations that determines the success of our fisheries. It is not our technological wood-cutting and processing capacity, but the amount of remaining forest land that affects the vibrancy of the forestry industry. The same goes for the relationship between the might of oil producers and the oil reserves under ground. This list can go on and on. As for man-made global warming as a key limiting factor among those mentioned above and already described by Daly several decades ago, the world economy has been feeling its effects at least since the beginning of the 21st century.

Much of the "prosperity" of the past three decades in the developed world, and particularly in the U.S., has been due to stock and real estate asset bubbles, rather than increases in real wages which have risen very slowly. An insidious cycle of borrowing to buy paper assets that drove up the price of paper assets created the illusion of prosperity. Many trillions of dollars were "created" in these markets. Millions of new homes were built and sold to investors who paid little or no money down, with the expectation that home prices would continue to rise 15%–20% per year. Many people went on a spending spree with their newfound wealth from growth of paper assets. Mortgages were packaged into structured investment vehicles of uncertain value that were marketed widely. Central banks, desperate not to "rain on the investors' picnic" nurtured the bubbles with loose monetary policies, driving up the pries of commodities, particularly oil When these bubbles expanded beyond the breaking point in late 2006–early 2007, the air began escaping from the bubble. It gradually became appaent that the illusion of wealth had been created worldwide as a result of bidding up paper assets to astronomical levels, far exceeding realistic assessments of growth of wealth based on gains in productivity and efficiency. And just as the apparent "value" of paper assets (mainly real estate and stocks) had grown to precipitous heights, they plunged at alarming rates in 2007–2008. As people no longer felt "wealthy", they ended their spending spree. This dragged down the price of commodities that now reversed their previous ascendance to unprecedented heights. The price of oil, which had previously climbed from $40/bbl to $150/bbl, fell back to $50/bbl in 2008. As of 2008, all indications were of a coming worldwide deflation and recession that could deepen into a depression. This has been observed with great alarm in the business and political circles of the developed world in view of a potential associated rise in unemployment. If all this points in the same direction, in the near future humanity will have no choice but to fundamentally reassess its accepted assumptions about economic practice.

We may be at the point when the "open" cycle of production, whereby resources are taken out of nature "for free" in exchange for a generous portion of waste, will have to be replaced with a self-contained cycle. This new, balanced approach to the interaction of the economy and the environment would involve strict limits based on the capacity of the biosphere. The very notion of progress would have to be redefined in such a scenario as humanity's norms of consumption and economic expectations change and as we reassess exactly what is meant by "the standard of living". This is the smallest price that we can pay for our previous thoughtless attitude toward nature. If, on the other hand, we keep moving mechanically along a path established and well-traveled in the past two to three centuries, the price will become immeasurably higher. The "buy now, pay later" attitude that led to the Great Depression of the 1930s has outlived itself on many levels, which indicates that the word "strategic" in "strategic development" must acquire a truly new meaning.

References

Blanko, Kh. A. (1995). "Tretie tysiacheletie," *Latinskaia Amerika* **9**, 4–14 [in Russian].

Daly, H. E. (1977). *Steady-State Economy: The Economy of Biophysical Equilibrium and Moral Growth*. San Francisco.

Danilov-Danil'yan, V. I.; K. S. Losev (2000). *Ekologicheskii vyzov i ustoichivoe razvitie*. Moscow: Progress-traditsia [in Russian].

Losev, K. S.; V. G. Gorshkov; K. Ya. Kondratyev; V. M. Kotlyakov; M. Ch. Zalikhanov; V. I. Danilov-Danil'yan; G. N. Golubev; I. T. Gavrilov; V. S. Reviatkin; V. F. Grakovich (1993). *Problemy ekologii Rossii*. Moscow: VINITI [in Russian].

Losev, K.S. (2001). *Ekologicheskie problemy i perspektivy ustoichivogo razvitia v Rossii v XXI veke*. Moscow: Kosmosinform [in Russian].

Matishov, G. G.; V. V. Denisov (1999). *Ecosystems and Biological Resources of Russian European Seas at the Turn of the 21st Century*. Murmansk: Marine Biological Institute.

Russkaia Atlantida (n.d.) Rybinsk: OOO "Format-print" [in Russian].

Smith, A. (1976). "An inquiry into the nature and causes of the wealth of nations," in R. H. Campbell and A. S. Skinner (eds.), *Glasgow Edition of the Works and Correspondence of Adam Smith*, Vol. 2. Oxford, U.K.: Oxford University Press.

Treachy, M.; F. Wiersema (1995). *The Discipline of Market Leaders: Choose Your Customers, Narrow Your Focus, Dominate the Market*. New York: Addison-Wesley.

Tvardovsky, A. T. (1970). *Za daliu dal'. Iz liriki tekh let.* Moscow: Sovetsky pisatel' [in Russian].

Veber, A. B. (1999). *Ustoichivoie razvitie kak sotsial'naia problema.* Moscow: Institute sociologii RAN [in Russian].

Weitzsäcker E. von; A. B. Lovins; L. M. Lovins (1997). *Factor Four Doubling Wealth—Halving Resource Use: A Report to the Club of Rome.* London: Earthscan.

6

The crisis of values as the main cause of the ecological challenge

Gaining independence or breaking away from nature?—Ancient roots of modern rationalism—Worldview revolution of the Renaissance—A mindset that leaves nature "out in the cold"—Priorities of commercial and monetary relations—Scientific and technical progress as the new religion of the 20th century—Tricks of the rational mind—"Nature first died in the human soul"—The psychology of short-term gain vs. tomorrow's interests—Destroying the destroyer

And so, at the beginning of the third millennium, humanity finds itself in a state of deep systemic crisis characterized by specific social, demographic, and economic aspects. However, these aspects pale in comparison to the ecological threat before us. Unfortunately, it is humanity that finds itself at the center of the global environmental degradation process that has by now been extended to virtually the entire planet. At fault is our consumerist attitude toward nature and the flawed mindset of a world conqueror.

As has already been noted, this psychology did not appear yesterday. The first step on the road to viewing ourselves as "lords of the Universe" placed above nature occurred in the Neolithic Revolution when prehistoric humans shifted from the hunting-gathering lifestyle (simple appropriation of nature's bounty) to agriculture (molding nature with human hands). And these hands have controlled our own destiny, so to speak, ever since.

Given our relative independence of natural forces, our former feeling of being at one with the plants and animals has been lost. Once we felt special, singled out from among all other beings, we were quick to retire our pagan gods, all of whom represented forces of nature. These deities imposed taboos on anything that threatened their green realm, but they no longer mattered and were replaced by new gods created in man's image and likeness. Nature lost its status as an object of veneration and turned into nothing more than our resource base.

However, even in pagan societies (i.e., the early civilizations of Antiquity), we already find the beginnings of this consumerist attitude toward the natural environment. For example, in the Fertile Crescent (Turkey and Mesopotamia) and Easter Island, deforestation led to the decline of agriculture and corresponding devastating sociopolitical consequences. As Jared Diamond (1999) puts it, "Fertile Crescent and eastern Mediterranean societies had the misfortune to arise in an ecologically fragile environment. They committed ecological suicide by destroying their own resource base" (p. 411). And V. Krasilov (1992) notes a recurring evolutionary pattern in this connection with respect to the civilizations of the Mediterranean Basin: "Population growth, resource depletion, expansion, militarization, totalitarianism, overstressed resources (the forests of Italy were used to build the Roman fleet), deteriorating human habitat, spiritual degradation, loss of inner energy, collapse of the State's system" (p. 14).

Francis Bacon's famous dictum "knowledge itself is power" (*ipsa scientia potestas est* from *Meditationes Sacræ. De Hæresibus*, 1597) would later sum up a notion that had began to form already in Antiquity: namely, that positive knowledge can be used to transform nature and adapt it to human needs. This power can simultaneously change and even revolutionize the social, as well as economic aspects of one's reality. Bacon, coming out of the Renaissance, represented the European attitude of his time. After the "darkness" of the early Middle Ages, when nature was still God's creation and therefore required a certain amount of human stewardship (in European legislation of the time there were several normative constraints in this connection), Greco-Roman rationalism was re-emerging on a new foundation. In contrast to the humility ethic of the previous period, the Renaissance launched the proud modern age that viewed nature in a new way.

Now, new modern nations were being born as various European states shifted their sociopolitical systems to market-based relations and, in some instances, to liberal civil societies. In intellectual terms, modernity essentially constituted a second revolution of worldviews after the Neolithic Revolution, and a huge contribution toward this change was made by seafarers, writers, architects, and painters, as well as thinkers, such as Descartes, Galileo, Newton, Adam Smith, and others.

In the context of the new ideology, the human being appears at the center of the Universe as its most glorious manifestation. We witness the growth of faith in the human potential, human power, and especially the might of human reason capable of transforming and reordering the word. To quote the great Russian 20th-century poet N. Zabolotsky (1958),

> "Two different worlds are owned by man:
> one that created us,
> the other which in every age
> we make as best we can" (from *Na zakate*, p. 299).

And indeed, the great geographical discoveries and the monumental achievements of scientific thought, the invention of printing, and the flowering of cities: everything appeared to confirm the undisputable superiority of the world made by man. At the same time, untamed or primeval nature was being pushed back in our consciousness, made secondary or perceived as a mere backdrop for our grandiose actions.[1]

It is not inconsequential that paintings by various masters of the Renaissance present nature mainly as a background, not necessarily devoid of picturesque charm, but fairly conventional and distant. As for the literary works of the period, as opposed to what is observed in the poetry of Antiquity, descriptions of nature are very scant (with the exception of the narratives coming from the conquest of the Americas). On the other hand, the Classical and Baroque periods set the stage for the flowering of landscape architecture, as well as garden and park design, as a high art form. Although this may have opened a small window on nature, this was still a vision of the natural world shaped and adapted to what humans wanted to see: a regular, trim, cultured realm which acted as a background for court feasts and entertainments. The stamp of this tailor-made version of nature can be seen on the canvasses of famous Flemish and Dutch painters. As for a non-anthropocentric view of "useless" or "wild" nature, that approach appeared much later in the Western world: toward the late 18th and early 19th centuries as Romanticism revolutionized our esthetics and worldview.

Although this new approach to nature was also highly conventional and "romanticized", it struck a chord in the hearts of many people who could now conceive of the natural environment as something in its own right and independent of any practical gain or human interests. As time went by, more and more painters (Constable and Turner in England, the Barbizon school in France, Levitan and N. Roerich in Russia) and authors (Melville, Thompson Seton, Thoreau, Stifter, the later Pasternak) drew their primary inspiration, as well as answers to many burning metaphysical questions, from nature.

However, art and reality are two different realms. In the real, daily life of the 19th century, for example, the attitude toward the natural world was hardly an improvement. The free market had established itself definitively as the main mode of social interaction and thinking about the world (i.e., money talked while nature was silent). When choosing between untouched, living nature and profit, the agents of the Industrial Revolution rarely hesitated in their pursuit of short-term goals. An area of century-old forest did not stand a chance when it stood in the way of construction projects. A few wagonloads of cement easily outweighed a bunch of "useless" trees.

The 20th century witnessed the apex and triumph of rationalism. Chekhov's cherry orchard (from the eponymous play of 1904), cut down to make room for lucrative summer villas, symbolized the new age with particularly glaring poignancy. Having convinced ourselves of our omnipotence, we were sure that rational organization and advanced technology would enable us to resolve any problem facing humanity. To quote Blanko (1995), "having faith in science as if it were a new religion, man decided

---

[1] It is paradoxical that, in spite of anthropocentrism, the Renaissance saw the demise of the geocentric Universe whereby the way was paved for viewing our planet as the modest satellite of an ordinary star in an ordinary Galaxy.

to establish his dominion and direct the development of nature and society on Earth according to his judgment".

And how could we resist the new faith? Only a hundred years passed between the first light bulb (1879) and the first mass-produced PC (1981). In that time, humanity invented the internal combustion engine and the cathode-ray picture tube, brought about the first nuclear reaction, and penetrated the mysteries of the genetic code. The Earth was laced with a network of high-speed railways and highways. A dense invisible cloud of electromagnetic radiation, which can be picked up even from space, enveloped the planet. Canals linked up three oceans, the mightiest rivers were dammed, while bridges and tunnels connected islands and continents.

Thus, the 20th century alone was enough to realize almost all of our most daring dreams and fantasies: from the waters of life (antibiotics and organ transplants) to flying carpets (airplanes) to the Tower of Babel (skyscrapers) to Jules Verne's imaginary voyage to the Moon, etc. The latter, notably the Apollo Program (conceived while President J. F. Kennedy was still in office) was so methodical and systematic that it resembled less an audacious challenge hurled at the heavens than a routine launching of a new icebreaker. This is the impression created by punctual, perfect, stage-by-stage implementation of this fantastic undertaking which was stunningly expensive (25 billion 1969 dollars, equivalent to 135 billion 2006 dollars). The first earthlings, N. Armstrong and M. Collins, landed on the Moon without a hitch.[2]

It seemed that nothing could stop this triumphant march of scientific and technological progress which kept opening up new horizons before mankind, providing undreamed-of levels of comfort, satisfying our most whimsical needs, and most importantly creating a sense of being firmly ensconced on Earth under a secure armored dome of civilization. In *A Gentleman from San Francisco*, a tale by Nobel Prize winner Ivan Bunin, even the Devil himself is awed by the sight of an ocean liner calmly slicing its way through stormy, wintry ocean waves.

The ship, called the *Atlantis*, is a spectacular multideck vessel that carries crowds of revelers partying in exquisite halls flooded with light. In the roomy hold, however, there is the dead body of an American returning home after a cruise cut short. This situation symbolizes the frightening aspect of the triumphant self-confidence at the base of our scientific and technological progress. Death and disaster are just underneath the feet of those celebrating their victory over nature on the upper decks:

> "But there, on the vast steamer, in its lighted halls shining with brilliance and marble, a noisy dancing party was going on, as usual. On the second and the third night there was again a ball—this time in mid-ocean, during a furious storm sweeping over the ocean, which roared like a funeral mass and rolled up mountainous seas fringed with mourning silvery foam. The Devil, who from

---

[2] Admittedly, there was a hitch later on, but it occurred after the successful flights of Apollo 11 and Apollo 12, when an explosion in the propulsion bay of Apollo 13 forced the crew to turn back. The lunar landing never took place, but fortunately, everyone made it back to Earth unharmed.

the rocks of Gibraltar, the stony gateway of two worlds, watched the ship vanish into night and storm, could hardly distinguish from behind the snow the innumerable fiery eyes of the ship. The Devil was as huge as a cliff, but the ship was even bigger, a many-storied, many-stacked giant, created by the arrogance of the New Man with the old heart. The blizzard battered the ship's rigging and its broad-necked stacks, whitened with snow, but it remained firm, majestic—and terrible. On its uppermost deck, amidst a snowy whirlwind there loomed up in loneliness the cozy, dimly lighted cabin, where, only half awake, the vessel's ponderous pilot reigned over its entire mass, bearing the semblance of a pagan idol. He heard the wailing moans and the furious screeching of the siren, choked by the storm, but the nearness of that which was behind the wall and which in the last account was incomprehensible to him, removed his fears. He was reassured by the thought of the large, armored cabin, which now and then was filled with mysterious rumbling sounds and with the dry creaking of blue fires, flaring up and exploding around a man with a metallic headpiece, who was eagerly catching the indistinct voices of the vessels that hailed him, hundreds of miles away. At the very bottom, in the under-water womb of the 'Atlantis,' the huge masses of tanks and various other machines, their steel parts shining dully, wheezed with steam and oozed hot water and oil; here was the gigantic kitchen, heated by hellish furnaces, where the motion of the vessel was being generated; here seethed those forces terrible in their concentration which were transmitted to the keel of the vessel, and into that endless round tunnel, which was lighted by electricity, and looked like a gigantic cannon barrel, where slowly, with a punctuality and certainty that crushes the human soul, a colossal shaft was revolving in its oily nest, like a living monster stretching in its lair. As for the middle part of the 'Atlantis,' its warm, luxurious cabins, dining-rooms, and halls, they radiated light and joy, were astir with a chattering smartly-dressed crowd, were filled with the fragrance of fresh flowers, and resounded with a string orchestra" (Bunin, pp. 56–57).

However, the contrast created by the Russian author is not only between the securely sheltered luxury and idleness of the dancing passengers, on the one hand, and the fury of the wintry ocean outside their steel and glass bubble, on the other. The hellish furnaces below the happy people consume something in addition to the "mountains of coal in their red-hot maws": the effort of human muscle "soaked in salty, dirty sweat and shining red from the flames." This slave-like labor is the force that maintains the irrepressible "feast of life" in the dance halls, lounges, and on decks. Such was the harsh reality beneath the hope and optimism of the young 20th century, at least as Bunin saw it.

And yet, already by the middle of the 20th century, progress eliminated the inhumane labor conditions that it had generated in the 19th century. For example, most of the coal-based fleet was scrapped and the outdated ships were replaced by more efficient, ergonomic turbo-vessels that ran on residual fuel oil. In the same manner the problem of child labor in the textile sector (depicted by Jack London in *The Apostate* at about the same time as when Bunin wrote his tale) was eliminated with the introduction of shuttle-less looms, again by the very industry that had created it.

And so it seemed that only time and patience were required before the arrival of general well-being would be brought to mankind by science and technology and inspired by the insatiable power of human inventiveness. However, this was only the surface view. The real price to pay lay hidden from the view of most people: the way Bunin's maritime revelers do not suspect that they are being watched by the Devil.

The Soviet Union was perhaps the most progress-crazed country on Earth. Here science and technology were more of a religion than anywhere else because the Workers' State was based on official atheism and a cult of human reason (which, however, did not prevent the medieval-like witch-hunts unleashed against Stalin's enemies and the population in general). In the 1920s to 1930s a whole generation of Soviet people lived in feverish expectation that Communism (the New Millennium) was just around the corner. For example, in *The Golden Calf*, an immensely popular novel written by I. Ilf and E. Petrov in 1931, we find the following vision of a technological future in Soviet Russia:

> "Perhaps Russian émigrés, driven to distraction by selling newspapers on the asphalt fields of Paris, recall the country roads of Russia with all the charming detail of their native landscape: the crescent moon lying in a puddle, the crickets praying loudly, and the clanking of empty buckets tied to peasant carts. But nowadays the light of the moon has another function in Russia. It will soon shine just as well on tarred roads. Motor car horns and klaxons will replace the symphonic music of the peasant bucket. And you will be able to hear the crickets in special sanctuaries."

This belief in the limitless possibilities offered by civilization was essentially not very different from the mindset of many Western intellectuals. Admittedly, by now some thinkers in the Western world were beginning to have their doubts, especially in the wake of WWI when science and technology had been used to massacre millions for no good reason. But there were still many optimists west of the Iron Curtain too. Thus, Western popular science literature occasionally demonstrated the same kind of indefatigable faith in a bright future and the psychology of "man the conqueror" who was sooner or later bound to tower above the entire natural world. For example, here is how the famous British astrophysicist Sir James Jeans ends his otherwise ideologically detached book entitled *The Universe around Us*, written roughly at the same time as *The Golden Calf*:

> "As inhabitants of the Earth, we are living at the very beginning of time. We have come into being in the fresh glory of the dawn, and a day of almost unthinkable length stretches before us with unimaginable opportunities for accomplishment. Our descendants of far-off ages ... will see our present age as the misty morning of the world's history; our contemporaries of to-day will appear as dim heroic figures who fought their way through jungles of ignorance, error and superstition to

discover the truth, to learn how to harness the forces of nature, and to make a world worthy for mankind to live in."[3]

Intellectual intoxication is no less fraught with danger than physical drunkenness since both prevent us from seeing reality as it is. And this reality was moving in a very different direction. The above-mentioned disaster of WWI, coupled with WWII, the two bloodiest conflicts in human history and both fueled by the fruits of scientific progress, were monstrosities of the 20th century. The same applies to the record-breaking cruelty of the century's totalitarian regimes: from Stalin to Hitler and from Mao Tse-Tung to Pol Pot, this phenomenon became something like the alter ego of our time. To this can be added the Holocaust and other similar genocides, as well as an endless stream of interethnic local wars that cost tens of millions of human lives. Finally, the cynical bloodletting by modern organized crime, such as the Italian Mafia and its post-Soviet Russian counterparts, as well as the outrages perpetrated by international terrorism in the last third of the 20th century, has also stamped our lives and perception of today's world.

To be fair, cruelty had been inseparable from the human condition throughout the ages, going all the way into the depths of prehistory. However, at least back then it was justified by the spiritual "darkness" of people whose sense of empathy for the other, a different tribe or religion, was severely limited by the culture of their times. But how can one explain the chain of murderous violence in the Age of Reason lit up by the powerful projector of advanced civilization? Perhaps sophisticated new ideologies, including faith in progress, made it possible to justify the "necessary". Progress and the bright future needed a clear path, and the rationalist logic allowed little room for compromise. The result was that age-old taboos and interdictions, which had already proven to be a rather imperfect bulwark against barbarity, could now be swept away by new demagogues with absolute confidence.

However, in the long run the most dangerous and destructive form of arrogance has turned out to be our utilitarian and consumerist attitude toward the natural environment, again made possible by the rationalistic giant with feet of clay. This is particularly starkly and cynically expressed by an aphorism expressed by Bazarov, the hero of I. Turgenev's novel *Fathers and Children*: "Nature's not a temple, but a workshop, and man's the workman in it" (ch. 9). It ought to be added that this "workman" appears to be interested in nothing but the short term, and is prepared to destroy the "workshop" for his goals.

At the same time, some of the best minds in the 19th century, unlike Bazarov, were perfectly capable of understanding the truth long before ecology appeared as a distinct discipline of inquiry. In A. Chekhov's play *Uncle Vanya*, Doctor Astrov tries

[3] Perhaps even more significant is F. D. Roosevelt's faith "infinite progress in the improvement of human life" (Roosevelt, 1941). In the same vein F.D.R. maintains that "among men of good will science and democracy together offer an ever-richer life and ever-larger satisfaction to the individual. With this change in our moral climate and our rediscovered ability to improve our economic order, we have set our feet upon the road of enduring progress" (Roosevelt, 1937).

to explain the situation in one of pre-revolutionary Russia's remote districts to Elena Andreevna, a St. Petersburg socialite. Astrov uses a home-made map:

> "Look there! That is a map of our country as it was fifty years ago. The green tints, both dark and light, represent forests. Half the map, as you see, is covered with it. Where the green is striped with red the forests were inhabited by elk and wild goats. Here on this lake, lived great flocks of swans and geese and ducks; as the old men say, there was a power of birds of every kind. Now they have vanished like a cloud. Beside the hamlets and villages, you see, I have dotted down here and there the various settlements, farms, hermits' caves, and water-mills. ... Now, look lower down. This is the country as it was twenty-five years ago. Only a third of the map is green now with forests. There are no goats left and no elk. The blue paint is lighter, and so on, and so on. Now we come to the third part; our country as it appears to-day. We still see spots of green, but not much. The elk, the swans, the black-cock have disappeared. It is, on the whole, the picture of a regular and slow decline which it will evidently only take about ten or fifteen more years to complete" (Act 3).

It is significant that Doctor Astrov is perfectly aware of the direct connection between the miserable state of the environment and the values of the people living next to it. The doctor's diagnosis is anything but medical as he completes his ecological speech in Act 3: "It is the consequence of the ignorance and unconsciousness of starving, shivering, sick humanity that, to save its children, instinctively snatches at everything that can warm it and still its hunger. So it destroys everything it can lay its hands on, without a thought for the morrow."

Doctor Astrov's dreams of a better life for his people run up against a wall in a way that may have seemed specific to Russia at the turn of the 20th century. However, the developed world, with its factories, highways, and railways, as well as its higher standard of living and general quality of life, was not substantially different in ecological terms from Chekhov's motherland. In the U.S.A. and Western Europe forest clearing and prairie plowing, as well as the elimination of wild animals and birds, were progressing even faster than in relatively backward Russia. As the new century continued, these countries came to the brink of wiping out their natural ecosystems completely. The result was the transformation of this huge region into one global zone of environmental destabilization. The Russian film actor and director Rolan Bykov said the following in the last television interview before his death: "Nature first died in the human soul and brain. It disappeared from the range of our purposeful activity and became unimportant." The problem is that nature has many ways of reminding us that it is by no means "unimportant".

One such reminder has to do with contact between human civilization and the animal world. As we appropriate or destroy 20%–40% of the vegetational biota's primary production, we are eating a huge number of other species out of house and home (Imhoff et al., 2004; Vitousek et al., 1986). At the same time a whole army of hangers-on (rats, crows, cockroaches, and similar animals) have adapted to a mode of existence at the expense of human beings. As our cities expand and swallow up increasingly more countryside, garbage-scavenging raccoons, skunks, and other displaced species have no choice but "move in" with us. By encouraging this parasitic life-style in animals that would otherwise fit perfectly into natural niches in the biosphere, we constantly impoverish and destabilize the environment. Having created artificial conditions for certain species, we are undercutting the natural selection process that has helped the biosphere survive and keep itself clean for almost 4 billion years in a constantly changing world.

However, what appears most unsettling for human interests is the link between human–animal contact and disease. In recent years a familiar scourge of mankind appears to have re-emerged: bubonic plague. And here urbanization may be a decisive factor. A few years ago Chanteau *et al.* (1998) reported a shift in the incidence of plague from rural areas of Madagascar to the capital city: "The plague's earlier comeback in the inland capital, Antananarivo, arose as city sprawl and shoddy housing put residents in closer contact with black rats" (Blue, 2008; Chanteau *et al.*, 1998). To this can be added the increase in cases of cholera, malaria, and dengue fever, which scientists have linked to the global warming caused by the "progress and science" that we have unleashed upon the world (Patz *et al.*, 2005). Furthermore, the rise of the Nipah virus in Malaysia and the worldwide proliferation of the West Nile virus are alarm bells that we cannot ignore. M. Levy, a global change expert at Columbia University's Earth Institute and part of a scientific team that recently reported on modern emerging animal-transmitted diseases in the journal *Nature*, sums up the danger as follows: "We are crowding wildlife into ever smaller areas, and human population is increasing. Where those two things meet, that is the recipe for something crossing over" (*BBC News*, February 20, 2008; Jones *et al.*, 2008).

Could this be the biosphere's way of "fighting back" against what we call progress? Perhaps there are global mechanisms of self-purification aimed at eliminating sources of biospheric destabilization. And if mankind is a destabilizing force, it is conceivable that such animal-mediated diseases as avian flu, Ebola and AIDS might also constitute nature's way of getting rid of the disruptor. So far the very science and technology that brought us to the current crisis have made it possible to hold back the tide, more or less, although the numbers of worldwide casualties of AIDS are hardly indicative of successful resistance. However, given current population, urbanization, deforestation, and pollution trends, the walls of the human fortress might crack, and then no amount of ingenuity will be enough to resist the fury of outraged Mother Nature.

References

BBC News, February 20, 2008. Available at *http://news.bbc.co.uk/2/hi/science/nature/ 7252923.stm*

Blue, L. (2008). "Return of the plague," *Time* (February 18).

Blanko, Kh. A. (1995). "Tretie tysiacheletie," *Latinskaia Amerika* **9**, 4–14 [in Russian].

Bunin, I. [1918]. *The Gentleman from San Francisco* (translated by A. Yarmolinsky). Boston: The Stratford Company.

Chanteau, S.; L. Ratsifasoamanana; B. Rasoamanana; L. Rahalison; J. Randriambelosoa; J. Roux; D. Rabeson (1998). "Plague, a reemerging disease in Madagascar," *Emerging Infectious Diseases* **4**(1), 101–104.

Chekhov, A. [1999]. *Uncle Vanya* (translated by Marian Fell). Project Gutenberg eBook 1756. Available at *http://www.gutenberg.org/dirs/etext99/vanya10h.htm*

Diamond, J. (1999). *Guns, Germs and Steel: The Fates of Human Societies*. New York: Norton, 1999.

Ilf, I.; E. Petrov (1931 [1962]). *The Golden Calf* (translated by J. H. C. Richardson). New York: Random House. See also *http://www.idlewords.com/telenok/6.htm*

Imhoff, M. L.; L. Bounoua; T. Ricketts; C. Loucks; R. Harriss; W. T. Lawrence (2004). "Global patterns in human consumption of net primary production," *Nature* **429**(24), June, 870–873.

Jeans, J. (1929). *The Universe around Us*. Cambridge, U.K.: Cambridge University Press.

Jones, K. E.; N. G. Patel; M. A. Levy; A. Storeygard; D. Balk; J. L. Gittleman; P. Daszak (2008). "Global trends in emerging infectious diseases," *Nature* **451**, 990–993.

Krasilov, V. A. (1992). *Okhrana prirody: printsipy, problemy, prioritety*. Moscow: Institut okhrany prirody i zapovednicheskogo dela [in Russian].

London, J. *The Apostate*.

Meditationes Sacrae. De Haeresibus (1597).

Patz, J.; D. Campbell-Lendrum; T. Holloway; J. A. Foley (2005). "Impact of regional climate change on human health," *Nature* **438**(17), 310–317.

Roosevelt, F. D. (1937). *The Second Inaugural Address*, January 20. Available at *http://www.bartleby.com/124/pres50.html*

Roosevelt, F. D. (1941). *The Third Inaugural Address*, January 20. Available at *http://www.bartleby.com/124/pres51.html*

Turgenev, I. [2000]. *Fathers and Children* (translated by Constance Garnett). New York: Bartleby.com. Available at *http://www.bartleby.com/319/2/9.html*

Vitousek, P. M.; P. R. Eriich; A. H. E. Eriich; P. A. Matson (1986). "Human appropriation of the product of photosynthesis," *Bioscience* **36**, 368–373.

Zabolotsky, N. A. (1989). *Stolbtsy i poemy. Stikhotvorenia*. Moscow: Khudozhestvennaia literatura [in Russian].

Part III

The world community: Politicians and scientists in search of a solution

7

The mission of the Club of Rome

New concerns in a post-war world—First environmental measures by the U.N.—The Club of Rome's specific features and status—Computer-predicted ecological collapse— The concept of the limitation of growth—Civilization beyond sustainability—Three models of a probable future

The first symptoms or harbingers of today's global crisis appeared in the first part of the 20th century. Some exceptional thinkers of the time (e.g., Nikolai Berdiaiev, Oswald Spengler, Erich Fromm, and others) repeatedly tried to make the public aware of various aspects of this problem. However, two vicious world wars ended up crushing the entire first half of the 20th century, which was followed by an exhausting standoff created by the Cold War. This turmoil displaced the interests and concerns of humanity away from what appeared back then to be "secondary issues". In the meantime, the economic boom following on the heels of World War II, as well as the concurrent arms race accompanied by the thoughtless waste of resources, led to even more barbaric environmental destruction. Now, however, the former colonial countries (the Third World) were part of the "dirty" equation. Furthermore, the post-war decades endured a series of atomic weapon tests in the atmosphere and the vigorous development of the chemical industry, including the production of polymeric materials, as well as a widespread expansion in the use of fertilizers and chemical pesticides in the agricultural sector. It is therefore no wonder that, as of the 1960s, the international community began to take notice of the environmental pollution problem and the need to protect wild nature.

In 1961, the Office for Economic and Social Council at the U.N. (OESC) adopted the historic Resolution Number 810 that dealt with the need to create a worldwide network of nature reserves and areas under special protection. Another important event in the environmentalist sphere took place that same year: the creation of the World Wildlife Fund for Nature (WWF). The fund began its activity by financing environmental protection on the Galapagos Islands, a unique natural monument, and later ended up making an important contribution to the preservation of many endangered biological species. Five years later (in 1966) the International Union for Conservation of Nature was instrumental in publishing the first International Red Book containing a detailed list of endangered flora and fauna in the world. Subsequently, similar books dealing with this issue in particular parts of the world were published in many countries, which paved the way for systematic efforts in this area.

Under the aegis of UNESCO, another highly significant undertaking was conceived and carried out in the same period: the International Biological Program (IBP). It took ten years to bring this about (1964–1974), and the result was a major contribution to the study of the structure and functional principles underlying the main types of ecosystems. In the framework of the International Biological Program, numerous land and sea expeditions were organized, which made it possible to inventory the remaining natural resources of the world and to carry out a general assessment of the biosphere's general state. The first results of this research were summed up at the General Conference of UNESCO in 1970 where it was agreed to adopt the international Man and Biosphere Program. The program was meant to attract the attention of the international scientific community to the problem of biospheric sustainability under conditions of anthropogenic pressure.

However, it was not only through the U.N. that the first serious steps were taken toward grasping the global ecological situation and the associated prospects for world development. An equally important contribution to the solution of this problem was made by the Club of Rome, an international non-government organization (NGO) that brought together politicians, business people, and scientists from all over the world (around 100 participants in total). These individuals were too concerned about the fate of the world to merely observe the symptoms of the approaching crisis without trying to find new paths toward development. The initiator of the club's creation was Aurelio Peccei (1908–1984), vice-president of the Olivetti Company, and the name of the organization came from the Italian capital where the club's founders met at the Accademia Nazionale dei Lincei in 1967 for the first time. The key purpose of the Club of Rome was to mobilize leading specialists in the field of technological forecasting in order to deal with the solution to the crisis via system analysis, taking into account interconnections among ecology, demographics, economics, and the resource sector. The research projects initiated by Peccei's group were financed by various large corporations and were concerned with various aspects of the planetary crisis. The international scientists involved presented their findings in the form of reports to the Club of Rome.

At that time, futurology began to rely more and more on quantitative mathematical methods, which explains why in 1970 the Club of Rome solicited the assistance of Jay Wright Forrester, professor at the Massachusetts Institute of Technology (MIT), who was the first to use computer modeling for investigating global development trends. However, Forrester preferred to delegate the job to his pupils: Donella H. Meadows (1941–2001) and Dennis L. Meadows (b. 1942). The Meadows and their collaborators from the field of computer science used improved computer models to calculate a series of projected world development scenarios for the years 1970–2100.

In this project the totality of factors affecting the development of production and population growth was divided into two categories: the *physical* (food, raw materials, nuclear fuel, capacity to absorb industrial pollution by surviving ecosystems) and the *social* (the stability of social structures, education, employment levels, etc.). It was established that by the middle of the 21st century, even the most optimistic of the projected scenarios amounted to inevitable ecological collapse against the background of a deteriorating biosphere (Meadows *et al.*, 1972). Although it was subsequently determined that the initial scenarios used by the researchers were not free of certain errors, this first report to the Club of Rome, entitled *The Limits to Growth*, had a truly shocking effect and was even perceived by some as a doomsday prediction; this despite the authors' own claim that the report dealt merely with a choice of possible future events, rather than the future itself.

The main achievement of this and subsequent reports, *Dynamics of Growth in a Finite World* (Meadows *et al.*, 1974) and *Mankind at the Turning Point* (Mesarovic and Pestel, 1974), was arousing the awareness of the broad public of a seemingly obvious truth: that the exponential economic growth spanning already numerous generations has limits that can be objectively set. These limits are determined by the very finite nature of our planet, particularly its non-renewable resources, and even renewable natural resources, because the latter's rate of renewal is several orders of magnitude slower than the rate of resource use. At the same time it was demonstrated that given unchanged current trends towards increasing production volumes and population, the limits of growth on Earth could be reached in the foreseeable and quite tangible future rather than at some distant point beyond the horizon. Already back in 1798, Thomas Robert Malthus had enough vision to warn about this in his work *An Essay on the Principle of Population*.

Two decades later the authors of the first two above-mentioned reports once again addressed the same problem in a book eloquently titled *Beyond the Limits* (Meadows *et al.*, 1992). Their conclusion was in a way a snapshot of the global situation that had formed after 20 years of world development in the framework of an outdated socio-economic paradigm. The authors argued that despite technological fine-tuning, improved scientific knowledge, and stricter environmental protection policies, our civilization had already gone beyond the limits of its own sustainability in many parameters, both with respect to resources and environmental pollution. Thus, humanity had essentially transgressed the limits of growth and was even then beyond the zone of sustainability.

Beyond the Limits formulates a set of seemingly simple questions that are, nonetheless, of vital importance for public awareness in the international community. How many people can our planet sustain? At what level of material consumption? For how long? How overstrained is the biosphere that supports the human race along with its economy, as well as other biological species? How flexible is this support? How many sources and kinds of stress can the biosphere protect us from?

The numerous computer-modeling experiments conducted by the authors of this research demonstrated yet again that, under conditions of unchecked growth, the biosphere's limits will be reached sooner or later. Although reaching any given limit could in principle be artificially "postponed" through the use of various economic mechanisms or newest technologies, in both cases constant economic growth guarantees collision with other, new limits. Furthermore, the more that humanity is successful at postponing the manifestation of a given limit, the more likely it becomes that we will face a number of limits simultaneously since these limits are coupled in a systemic relationship.

Unfortunately, having correctly dealt with the critical state of the biosphere as an integral self-organizing system, the authors of *Beyond the Limits* seem to have neglected the system aspects and used only aspects of environmental pollution in their computer models. As a result, the destruction of natural ecosystems and that of the wild biota making up these ecosystems is a problem that was left out of the equation. In other words, the models in question approached nature as an ordinary resource from the same category as mineral resources.

Only when *Beyond the Limits* deals with the scale of humanity's usurpation of the primary photosynthetic product, do we glimpse a truly profound understanding that the fate of civilizations and the future of the Earth's biota are inseparable: "Somewhere along the path of NPP [net primary production] usurpation, there lie limits. Long before the ultimate limits are reached, the human race becomes economically, scientifically, aesthetically, and morally impoverished" (p. 66). The authors add that "the future, to be viable at all, must be one of drawing back, easing down, healing" (preface, p. xv).

This book, which is remarkable in its own way, sums up the situation with the following futuristic conclusion: "The world faces not a preordained future, but a choice. The choice is between models. One model says that this finite world for all practical purposes has no limits. Choosing that model will take us even further beyond the limits and, we believe, to collapse. Another model says that the limits are real and close, and that there is not enough time, and that people cannot be moderate or responsible or compassionate. That model is self-fulfilling. If the world chooses to believe it, the world will get to be right, and the result will also be collapse. A third model says that the limits are real and close, and that there is just exactly enough time, with no time to waste. There is just exactly enough energy, enough material, enough money, enough environmental resilience, and enough human virtue to bring about a revolution to a better world. That model might be wrong. All the evidence we have seen, from the world data to the global computer models, suggests that it might be right. There is no way of knowing for sure, other than to try it" (p. 236).

And so, it took the authors 20 years to go from a relatively meek idea formulated in *The Limits to Growth* (1972), namely that choosing a future is possible, to the conclusion that an adequate choice implies a radical change in world development. Approximately the same amount of time was required for the international community to grasp the urgency of the situation, which is the subject of the next chapter.

References

Malthus, T. R. (1798). *An Essay on the Principle of Population.*

Meadows, D. H.; D. L. Meadows; J. Randers; W. W. Behrens III (1972). *The Limits to Growth: A Report for the Club of Rome's Project on the Predicament of Mankind.* Universe Books, New York.

Meadows, D. L.; W. W. Behrens III; D. H. Meadows; R. F. Niall *et al.* (1974). *Dynamics of Growth in a Finite World.* Wright–Allen Press, Cambridge, MA.

Meadows, D. H.; D. L. Meadows; J. Randers (1992). *Beyond the Limits.* Chelsea Green Publishing, White River Junction, VT.

Mesarovic, M.; E. Pestel (1974). *Mankind at the Turning Point: The Second Report to the Club of Rome.*

8

Programs for change:
Stockholm–Rio–Johannesburg

The shift of ecological concerns to the international realm—The Stockholm Conference and its indirect consequences—The formation of an ecological consciousness—The Brundtland Commission as a stepping stone on the road to Rio—The birth of the term "sustainable development"—What was left unsaid by the Rio Summit—Rio + 10: the summit of unfulfilled hope—The disturbing results of the past 15 years—Is humanity prepared for change?

In 1968 the U.N. General Assembly decided to convene a meeting in Stockholm for the year 1972. This became known as the United Nations Conference on Human Environment (or "The Stockholm Conference"), and its purpose was to discuss a whole series of problems that had caused profound concerns in the international community. These included the growing threat of nuclear conflict; the continuing arms race; and the symptoms of environmental deterioration brought about by constantly increasing production and consumption, as well as accelerating population growth on the planet. The conference was preceded by the International Seminar on Development and Environment that took place in 1971, in a small Swiss resort town called Thun. This was destined to become the first way station of its kind in the quest for global sustainability. For the first time, the general ecological threat was addressed, and attention was given to the significance of the problem for the Third World. However, the key impact of the seminar was to prepare the groundwork for the United Nations Conference on the Human Environment that came together in Stockholm on June 5–16, 1972 (cf. UNEP, 1972a).

The conference publicly confirmed what was already a continuing cause for alarm in the scientific circles: the serious ecological problems affecting not only specific regions, but the planet as a whole. As in the case of the reports to the Club of Rome, it was demonstrated that the environment should not be viewed in isolation from the development of civilization. The participants at the conference recognized that the direction of world development as a means of satisfying growing human needs had come into serious conflict with the needs to preserve the environment. This point was demonstrated in the form of models created for the reports to the Club of Rome, as well as contemporary scientific observations, including satellite data and objective quantitative criteria.

Furthermore, the Stockholm Conference adopted a declaration consisting of 26 principles, recommended as guidelines for the international community. This document was the first such publication to address issues, such as peaceful co-existence, the economic underdevelopment of the Third World, social inequality, and the planet's ecological problems, in a comprehensive, systemic manner. In contrast to dealing with such typical Cold War issues as ideological or military confrontations, the declaration placed in the foreground the need to preserve and improve the environment in which we live and the related interests of the present, as well as future generations: "A point has been reached in history when we must shape our actions throughout the world with a more prudent care for their environmental consequences" (UNEP, 1972b).

Although the documents and decisions stemming from the Stockholm Conference had no legal weight and did not propose any ratification procedures on the part of individual states, their resonance was so great that they laid the foundation for a wide network of national environmental protection institutions and provided a powerful impetus for the development of environmental legislation. This period was also marked by the appearance of the so-called "green" social movement that spread over many countries, including the famous Greenpeace organization, as well as other less well-known environmentalist NGOs. A direct result of the Stockholm Conference, of primary importance was the formation of the United Nations Environment Program (UNEP) and a body created for its implementation, the permanently active UNEP Governing Council (a U.N. department for the environment) located in Nairobi, Kenya.

In this manner, as of 1972, environmental protection became a truly large-scale activity primarily aimed at the struggle against pollution. In the next 20 years, direct worldwide expenditures for this purpose alone amounted to approximately $1.5 trillion (Danilov-Danil'yan and Losev, 2000). Industrialized nations invested huge sums for the modification of so-called "dirty" technologies and nuclear energy, the latter being perceived at the time as ecologically cleaner and quite safe. Even after the events of the Chernobyl disaster (1986), many countries (France, Belgium, Finland, Japan, etc.) still prefer nuclear power, using it to provide 30%–70% of their electric power needs (Menshikov, 1998).

However, these were disparate and uncoordinated efforts toward environmental protection that could not decisively affect dangerous trends in world development. It became increasingly apparent that a worldwide program of action was needed. Such a program was in fact created, and the first step in this direction was made in 1983 by the World Commission on Environment and Development (WCED) formed under the auspices of the U.N. This body became part of history as the Brundtland Commission, so named after its chair, the Norwegian politician Gro Harlem Brundtland, who subsequently served as the prime minister of Norway.

The functions of the commission included the preparation of proposals for long-term environmental protection strategies and the elaboration of goals to be used as guidelines for working out practical policies by individual states. In 1987 the program report of the Brundtland Commission, involving the work of many international experts, was published under the title *Our Common Future* (U.N., 1987). This was soon followed by translations of the document into the principal world languages.

Without actually using the word "crisis", the authors of the report nevertheless ended up describing the state of the biosphere implicitly as a crisis. This was also the tone used in the discussion of the planet's demographic situation. However, having recognized the necessity of specific limitations in the area of natural resource exploitation, the authors deemed these constraints to be relative rather than absolute (i.e., determined by the level of technological development and existing social relationships). The latter two parameters, as it was argued, had to be controlled and perfected, which was supposed to open up possibilities for a new age of economic growth. The dubious nature of this position will be discussed below, but for now it ought to be stressed that the report did not provide an adequate assessment of the process of the disappearance of natural ecosystems. The biota was essentially viewed as an economic resource, admittedly endowed with ethical, esthetic, and cultural value (and this was by no means the first time this viewpoint was taken by a world body).

Although the Brundtland Commission was unwilling to proclaim the full scale of the ecological crisis in 1987, this was not the case in 1991, when a group of leading economists and ecologists published a book through UNESCO entitled *Environmentally Sustainable Economic Development: Building on Brundtland* (Goodland *et al.*, 1991). Edited by Robert Goodland, Herman Daly, Salah El Serafy, and Bernd von Droste, this publication had no compunctions in "calling a spade a spade". In particular, it was demonstrated that the global ecosystem was serving as an outlet for contaminants created by the economic subsystem. The latter had grown too big in relation to the biosphere, placing the capacity of biospheric sources and discharge channels under stress. Thus, until recently, humanity had been able to proceed with its economic activities without concern for the adaptive capability of the biosphere, viewing the world as something like a bottomless reservoir able to swallow up any amount of waste. However, now the "vacuous world" age had been superseded by the "filled world" period (Goodland *et al.*, 1991).

And yet, unfortunately, this serious and very profound piece of research failed to deal with the role of the biota in the formation of the environment and the allowable limits of its destabilization.

Both the *Brundtland Commission Report* and *Building on Brundtland* were on the table at the United Nations Conference on Environment and Development (also known as the Earth Summit, the Rio Summit, or UNCED) in Rio de Janeiro on June 3–14 of 1992. The scale of this conference was unprecedented and truly planetary, comprising participants from 178 countries, 114 heads of state, and representatives of 1,600 NGOs, as well as a huge number of journalists. Simultaneously, Rio hosted the Global Forum on ecological issues involving around 9,000 organizations, 29,000 individual participants, and 450,000 visitors who attended on their initiative. The Global Forum conducted a total of approximately 1,000 meetings and diverse events (U.N., 1992).

The greatest achievement of this conference was the process of open-minded intellectual engagement that accompanied the meeting. The wide discussion and exchange of opinions led to the general recognition of a key strategic premise: development and environmental problems can no longer be viewed separately. A convincing argument was made for the organic interdependence of poverty and underdevelopment in much of the Third World, on the one hand, and ecological issues, on the other hand. The faulty production and consumption system in the majority of the developed nations was recognized as the other key element in this equation. Pressure exerted by a growing population on nature, energy use, climate change, trade in tropical forest products, and desertification; all these aspects of global and regional ecology were discussed on an unprecedented scale and level, attracting more attention than ever before.

A number of documents were adopted in the course of the Earth Summit. The central ones were the *Rio Declaration on Environment and Development* and *Agenda 21*; these documents offered a plan of action for the attainment of ecologically sustainable development. The Rio Declaration reflected the evolution of worldwide ecological perceptions in the 20 years that had elapsed since the Stockholm Conference. This took the shape of principles proposed by the authors of the declaration as guidelines for the international community. Thus, Principle 1 postulated the leading role of the people or the general population (as opposed to only state structures) in the course of sustainable development. According to Principle 2, the state is seen as a guarantor of appropriate environmental quality and bears international responsibility for any harm caused to the environment. Special stress was placed by Principles 3–5 on the inseparability of goals underlying socio-economic development, including the struggle against poverty, as well as the environmental interests of the present and future generations. In Principle 10, great importance was attributed to the participation of the community in the elaboration of solutions to ecological problems, while Principle 11 was concerned with developing ecological legislation.

A special section of the declaration (Principle 15) was devoted to the precautionary approach: "Where there are threats of serious or irreversible damage, lack of full scientific certainty shall not be used as a reason for postponing cost-effective measures to prevent environmental degradation" (UNEP, 1992). Furthermore, it was recommended in Principle 16 that individual states rely on economic mechanisms of environmental protection and compensation for pollution. Principle 17 advocated the use of ecological impact assessment for evaluating the effect of planned activities on the environment; and in Principle 18 states were called upon to inform other countries of natural and other disasters with transnational consequences (UNEP, 1992).

Agenda 21, spread over more than 1,000 pages of text and based on the Brundtland Commission report, was the other key document of the Earth Summit. Analysis of the Stockholm plan of action demonstrates that most of its recommendations had to do with five problems: environmental assessment; environmental management;

determination and control of global pollution; ecological education, culture and information; and development and the environment. The new plan of action for the 21st century in *Agenda 21* shifted the focus toward questions of development, justice, and international cooperation. Therefore, the main sections of *Agenda 21* were placed in the following thematic order: (1) social and economic aspects (including demographic problems); (2) maintenance and rational use of natural resources for development purposes; (3) reinforcement of roles played by main population groups; and (4) means of implementation which, in addition to financial parameters, include scientific development, education, and international cooperation.

Agenda 21 included over 100 programs encompassing a very wide range of issues, from overcoming poverty to increasing the role of the community in the solution of ecological problems. However, certain important questions, such as demography, the structure of consumption, Third World debt and the export of dangerous waste, were unfortunately either underrepresented or completely absent from the document. Nevertheless, *Agenda 21* became an example to follow for national programs on the transition to sustainable development. The United Nations Conference on Environment and Development recommended such programs for every country in the world, and today more than 100 states have followed this recommendation.

Perhaps the most palpable result of the conference was the introduction and wide circulation of the term *sustainable development*. This concept was put forth as an alternative to the previous ecologically destructive course of civilization. The concept of sustainable development proposed by the Rio Summit was based to some extent on the report of the Brundtland Commission and included the following main points:

- Priority should be accorded to the people, who have the right to live a healthy and fruitful life in harmony with nature.
- Environmental protection should become an inseparable element in the process of development and cannot be viewed in isolation from the latter.
- The satisfaction of needs pertaining to the development and preservation of the environment must apply not only to the present but also to future generations.
- Among the most important goals of the international community are the diminution of the gap in the standard of living between different countries and the eradication of poverty.
- In order to attain sustainable development, states must eliminate or decrease the role of production and consumption models that do not favor this goal (Shlikhter, 1996).

However, one must not overlook the criticism that can be leveled against some of UNCED's positions that at times appear to contradict one another. Thus, one of their points allows for the possibility of further civilization growth based on the notion of the above-mentioned "vacuous world" model. However, we read elsewhere about the need to limit consumption and reduce the planet's population on the basis of the "filled world" concept: that is, a world close to having reached its ecological limits, etc. (Danilov-Danil'yan and Losev, 2000).

Such contradictions will be dealt with below, but for now we need to address the generally unsatisfactory aspects of certain documents summing up the breakthrough conference in Rio. To begin with, even though we witness the admission of global environmental change (i.e., deforestation, the decrease in biodiversity, dangerous climate change, etc.) not one of the documents admitted the fact that the planet had already entered a phase of full-scale ecological crisis or that this crisis requires a radical reassessment of existing principles governing world development. Most importantly, there was no attempt to elaborate a scientifically based strategy for such development, a strategy that would require a serious theoretical foundation.

Instead, most of the conference participants appeared to rely on pure empiricism and experience accumulated in the past to arrive at their understanding of the problems at hand. It was believed that this experience had demonstrated the monumental capacity of the human race to find solutions in the tightest of tight spots by means of scientific and technological achievements or the improvement of socio-economic institutions. Counting on such seemingly tried and tested tools, the conference participants attempted to apply this wisdom to problems that civilization had never faced before. The proposed methods involved the structural and technological reconstruction of industry, the use of low-pollution technologies, etc.

Meanwhile, in the second half of the 20th century, humanity had made a discovery of such enormous proportions that it became impossible to retain lines of thought that had been perceived as acceptable in the past. That discovery was the environment. Having ignored it for many centuries as something marginal and only indirectly related to our lives, humankind suddenly realized that the environment was in fact directly linked to every single aspect of human existence: from the world economy to the state of our health. Humanity was now aware that the environment functioned according to its own mechanisms and imminent laws, and the study of these laws should have started about a century earlier. This oversight had led to a confrontation with nature, and resolving this conflict after long neglect would be very difficult indeed. By following certain outdated notions and habits, the participants of the Earth Summit had overlooked some of the new concepts stemming from the recently acquired ecological insights.

In accordance with decisions made in Rio, the next world forum on the environment was supposed to be held ten years later. This event took three years to prepare, and it was opened in August of 2002 in Johannesburg, South Africa. Both the numbers of participants and their representative importance almost matched the scale of the Rio conference (*http://www.un.org/jsummit/*). However, there was one major difference. While expectations all over the world had been great in the case of the Earth Summit, unfortunately the Johannesburg Summit (also known as Rio + 10) did not have this advantage. This low level of enthusiasm was attributable in part to the insignificant progress achieved in the area of sustainable development in the previous ten years. The disappointment was more than justified:

> "... progress in implementing sustainable development has been extremely disappointing since the 1992 Earth Summit, with poverty deepening and environ-

mental degradation worsening. What the world wanted, the General Assembly said, was not a new philosophical or political debate but rather, a summit of actions and results" (U.N., 2002a).

"The global environment continues to suffer. Loss of biodiversity continues, fish stocks continue to be depleted, desertification claims more and more fertile land, the adverse effects of climate change are already evident, natural disasters are more frequent and more devastating, developing countries more vulnerable, and air, water, and marine pollution continue to rob millions of a decent life" (U.N., 2002b, p. 13).

Two key documents were adopted at the Johannesburg Summit: the declaration (cited above) and the *Plan of Implementation of the World Summit on Sustainable Development*. Both documents placed poverty at the top of their agendas, viewing poverty as the key factor in social instability, criminality, and moral dysfunctionality. Poverty can indeed create a sense of hopelessness and apathy leading to an attitude of irresponsibility to one's environment, both natural and social. Even one's own children can be victims of this neglect as they find themselves at the bottom of society and acquire a distorted notion of the world around them. The antisocial cycle initiated in this manner is very difficult to break.

Poverty, inseparable from the weak economies of various countries, is very much a function of unemployment. The latter constitutes one of the most malignant conditions in a given population. And the next element in this chain is education because most jobs in today's world require qualified personnel. Therefore, if there is any chance of breaking the vicious circle, accessible education appears to be the key. This is why the Johannesburg Declaration and the Plan of Implementation propose that states develop national programs for allowing the poor strata of their respective societies to have more access to education, as well as production resources, credit, and equal social opportunities on the labor market (Marfenin, 2006).

The Plan of Implementation also contains a number of other important recommendations with respect to the social and ecological domains. For example, it deals with the need to give the poor access to agricultural resources and the free transfer of methods, as well as know-how pertaining to sustainable agriculture. The developing world must be given technologies ensuring the supply of cheap energy (e.g., sources of bioenergy, wind generators, and small-scale hydroelectric stations, etc.). The same goes for the development of agricultural methods for the prevention of soil and water resource degradation (U.N., 2002c). The Johannesburg Summit also recognized the need for fundamental changes in existing production and consumption systems, calling upon different countries to encourage sustainable models that would neither harm the environment nor undermine the natural resource base. For the first time at such a high-level meeting attention was given to the controversial problem of globalization with its positive and negative consequences for various countries and regions.

However, while in the course of preparation for and actuation of the Rio Summit the first steps (however timid) were made toward elaboration of a sustainable development concept, the Johannesburg Summit preferred to avoid consideration of such cardinal questions. Stress was placed on individual current problems, which were undoubtedly important and very pertinent (e.g., fresh water deficit, the food supply, energy, the preservation of biodiversity, etc.). In fact, the Rio Summit's entire approach appears to demonstrate that in order to deal with a whole series of vital problems faced by humanity, consistent concrete and practical measures were required. Not that such steps are unwelcome—in fact, they can improve the lives of millions—but they tend to merely patch up the wound instead of actually healing it. To quote Nitin Desai, the Secretary-General of the Johannesburg Summit, "We knew from the beginning of the Johannesburg process that the Summit would not produce any new treaties or any single momentous breakthrough" (U.N., 2002a).

Indeed, many of the agreed-upon target indicators had been adopted earlier during meetings at a lower level while most of the attention at the summit was given to various practical steps, as well as the coordination of multifarious goals, graphs, and obligations. As Nitin Desai put it, "I know some may have wanted more, but fulfilling these commitments will require new and additional resources" (U.N., 2002a). And so, we see an admission from the highest level that the Johannesburg Summit did not manage to fully tackle the fundamental challenge before us. The very breakthroughs that were not made (according to the above-cited statement by the secretary-general) are long overdue, and, if not breakthroughs, then at least the elaboration of a further strategic development plan. The key shortcoming admitted by the Johannesburg Declaration itself was that instead of moving closer to sustainability, the world was actually getting farther away from it. Attempting to solve various problems in isolation, without taking their profound and complex causes into account or looking at the connections between them, is most likely a completely pointless undertaking.

What complicates the issue still further is that the Earth summits do not take place every year or even every five years. Such meetings are of world-level significance, and so the whole world expects monumental outcomes from these events. When the question is "to be or not to be", while most indicators for the global environment keep pointing toward deterioration, half-measures simply will not do. How is this possible, despite the influence of the Stockholm, Rio, and Johannesburg Summits, the existence of UNEP and the fact that in all states larger than Honduras (except for Russia) there are functioning ministries or departments of the environment? Our failure is all the more puzzling given the volume of the ecology market that is now approaching the trillion-dollar mark per year.

If we turn to the question of long-term prospects, the two key failures of the three Earth summits are the persistence of poverty and the deterioration of the biosphere. The world has not managed to narrow the growing gap between the most and least developed nations, nor has hunger been eradicated on the planet. The associated demographic problem is also still there. As for the biosphere, the concentration of greenhouse gases is still increasing while forest-covered areas (excluding Europe) keep on shrinking. To this we can add the further expansion of deserts, the increasing loss of fresh water resources, the decrease of biodiversity, and so on. Perhaps the only success story that can provide a sense of satisfaction to the international community is the stabilization of the ozone layer. This has been achieved thanks to measures adopted by the Vienna Convention in 1985 (in force as of 1988) and the Montreal Protocol of 1987 (in force as of 1989) (Danilov-Danil'yan, 2007; UNESCAP, n.d.).

Of course, one should not underestimate the significance of this success story. After all, apart from direct or "physical" results, humanity has produced a convincing confirmation of a process that can be applied to much of what we have been discussing: that global ecological processes can be carried forward from theory to practice. However, this single triumph remains unique, an eloquent reminder of what could be achieved in the ecological sphere but has not taken place (i.e., the movement toward sustainability). Much of this is due to the noticeable cooling of interest in the issue of sustainable development that peaked in the 1990s. However, we are not arguing that the average person no longer cares about questions of ecological safety, environmental protection, or pollution. The problem is that these questions appear to exist on their own, as it were, while sustainable development is left out of the bigger picture, so that the interrelationship between the many elements of the puzzle receives little attention.

Furthermore, the media often misuse the term "sustainable development" in ways that have become fashionable and very widespread. Compounding the problem even more is the fact that after 9/11, and after the start of the Afghanistan and Iraq wars, the number of scholarly publications on questions of global environmental change has decreased significantly. Apparently, the international community has difficulty concentrating on everything at the same time, and what seems most pressing today ends up usurping the major attention. Furthermore, resources are not unlimited for research and the dissemination of information.

At the same time it is clear that humanity still appears unwilling to curb its consumerist appetite either in terms of quantity or variety. In fact, not only do we continue to desire to consume more, but there continue to be more and more consumers of everything, from plastic toys to fossil fuels. Although the percentage rate of population growth in the world as a whole has decreased, the population of the developing world is still growing at a significant pace. By the end of the current decade, 85% of the world's population will reside in developing countries where sociopolitical and economic obstacles sometimes make it all the more difficult to resolve the problems under discussion. This is another sign of the profound crisis in which modern civilization finds itself, in conflict with an environment whose sustainability laws impose limitations on the numbers and behaviors of all species, including ours of course.

References

Danilov-Danil'yan, V. I. (2007) "Sustainable development: 20 years of arguing." In: *Economic Effectiveness of Russia's Development*. Moscow: TEIS, pp. 467–480 [in Russian].

Danilov-Danil'yan, V. I.; K. S. Losev, (2000). *The Ecological Challenge and Sustainable Development*. Moscow: Progress-Traditsia [in Russian].

Goodland, R.; H. Daly; S. El Serafy; B. von Droste (eds.) (1991). *Environmentally Sustainable Economic Development: Building on Brundtland*. Paris: UNESCO, 1991.

Marfenin, N. N. (2006). *The Sustainable Development of Humanity.* Moscow: Izdatel'stvo MGU, [in Russian].

Menshikov, V.F. (1998). *Russia with or without Atomic Energy: Russia in the Environment* (analytical annual). Moscow: Izdatel'stvo MNEPU [in Russian].

Shlikhter, S. B. (1996). "The contradiction of sustainable development and problems inherent in overcoming them," *Geographic Problems Both Environmental and Social.* Moscow: Sovet po fundamental'nym geograficheskim problemam, pp. 59- 68 [in Russian].

U.N. (1987). *Report of the World Commission on Environment and Development: Our Common Future.* New York: United Nations. Available at *http://www.un-documents.net/wced-ocf.htm*

U.N. (1992). *United Nations Conference on Environment and Development.* Available at *http://www.un.org/geninfo/bp/enviro.html*

U.N. (2002a). *The Johannesburg Summit Test: What Will Change?* Available at *http://www.un.org/jsummit/html/whats_new/feature_story41.html*

U.N. (2002b). *Johannesburg Declaration on Sustainable Development: From Our Origins to the Future.* Available at *http://www.un.org/esa/sustdev/documents/WSSD_POI_PD/English/POI_PD.htm*

U.N. (2002c). *Plan of Implementation of the World Summit on Sustainable Development.* Available at *http://www.un.org/esa/sustdev/documents/WSSD_POI_PD/English/WSSD_Plan Impl.pdf*

UNEP (1972a). *Report of the United Nations Conference on the Human Environment.* Available at *http://www.unep.org/Documents.Multilingual/Default.asp?DocumentID = 97*

UNEP (1972b). *Declaration of the United Nations Conference on the Human Environment.* Available at *http://www.unep.org/Documents.multilingual/Default.asp?DocumentID = 97& ArticleID = 1503*

UNEP (1992c). *Rio Declaration on Environment and Development.* Available at *http://www.unep.org/Documents.Multilingual/Default.asp?DocumentID = 78&ArticleID = 1163*

UNESCAP (n.d.). *ESCAP Virtual Conference.* United Nations Economic and Social Commission for Asia and the Pacific. Available at *http://www.unescap.org/drpad/vc/orientation/legal/3_vienna.htm# Montreal*

9

Toward a systemic understanding of the biosphere

Influence of life on chemical state of the planet—Ecosystems—Biogeocenosis—The biosphere as a living factory—The Gaia Hypothesis—The Haeckel and Vernadsky heritages

Like a worm eating away at a favorite fruit from the inside, humanity has been constructing its civilization within the biosphere by partially destroying the latter. And only recently have we started to study this extremely complex system, although the first attempts to achieve a universal, holistic approach go back to the time of Alexander von Humboldt (1769–1859). One of von Humboldt's achievements was the elaboration of an alternative to the mosaic of independently existing species proposed by the Swedish botanist and zoologist Carl Linnaeus (1707–1778). That alternative was the notion of interaction or interconnectedness among organisms and the landscape around them. This constituted the foundations of biogeography, a science that views the climate as the determining factor in a given landscape. However, in the second half of the 19th century, von Humboldt's concept of a unified Earth system yielded the stage to phylogeny, the history of species origins and development, as the only noteworthy scientific understanding of natural phenomena (Zavarzin, 2004).

Charles Darwin (1809–1882) explained the Linnean diversity of species through phylogeny in the context of competition-based natural selection determined by the mutability and hereditary transmission of acquired characteristics. Darwin's breakthrough *Theory of Evolution* was as logical as it was independent of the need to postulate any external forces for the attainment of a biologically sustainable state. In fact, this approach ended up becoming not merely a theory—but an entire worldview. As the *Theory of Evolution* developed, reductionism prevailed in the biological sciences (i.e., the explanation of the general through the individual on the basis of accumulated empirical material). Scientists focused on the evolutionary lot of a given species or a single organism, which unfortunately led to conceptual biotic fragmentation. This tendency became an absolute and greatly slowed down the development of other approaches based on viewing the biosphere as a single system governed by all-encompassing laws and mechanisms. The latter position with respect to biospheric research was adopted by only a few daring thinkers at the turn of the 20th century.

One might expect that the systemic understanding of the biosphere would have formed within the framework of ecology that was beginning to take on its embryonic shape at the time. However, everything turned out differently. The first person to have independently reached a modern systemic position was not a biologist but a mineralogist, the founder of biogeochemistry V. I. Vernadsky (1863–1945). In 1926, this outstanding Russian scientist published a series of lectures under the general title: *The Biosphere* which appeared three years later in French translation entitled *La Biosphère*. Here, Vernadsky put forth the idea of a holistic and interdependent world of living matter whose living substance he called "biofilm". This system was unified by a network of biogeochemical cycles within abiotic spheres: the atmosphere, the hydrosphere, and the lithosphere. This circumterrestrial layer where biochemical processes take place was referred to by Vernadsky as the biosphere.

Vernadsky demonstrated that the chemical state of the planet's external shell is entirely under the influence of life (i.e., living organisms). His biospheric teaching considered not only the main properties of living substance and the way that non-living nature influences that substance, but also the fundamental importance of the reverse relationship (i.e., the influence of life on the abiotic environment and the resulting formation of non-living natural substances, such as the soil, that constitute habitats for living things). For the first time the whole living layer around the Earth appeared as a unified, complex and simultaneously fragile entity. Here is how Vernadsky summed up his ideas: "What the biosphere of our planet contains is not life distinct from its environment, but rather a living substance, i.e., a totality of living organisms intricately connected with the biospheric environment around it. This powerful geological factor is inseparable from the biosphere" (Vernadsky, 1987, p. 269). Vernadsky was also the first to formulate the idea that "thanks to the on-going and never-ending evolution of the species, sharp changes occur in the environmental impact of the living substance. As a result, the evolutionary process (changes) is transferred to natural non-living, as well as biogenic, bodies that play a key role in the biosphere. It is also transferred to the soil, underground and above-ground water (seas, lakes, rivers, etc.), coal, bitumen, limestone, organogenic ores, etc." (Vernadsky, 1988, p. 27).

However, pondering the special place that humanity occupies in the biosphere and the manner of the latter's evolution, Vernadsky went a step further by considering the possibility of possessing and controlling the biosphere through the power of the human mind:

"After grappling unconsciously with nature for many hundreds of thousands of years, the only social species of the animal kingdom—man—has completed the conquest of the biosphere's entire surface. Not a single corner of the planet remains inaccessible to mankind, and there are no limits to human proliferation. Scientific thought, as well as technology organized and directed by State structures, has made it possible for human beings to create a new biogenic force in the biosphere. This force directs our proliferation and ensures favorable conditions for settling those parts of the biosphere where human life (and in some places— life of any kind) could not exist in the past. In theory we see no limits to the human potential" (Vernadsky, 1988, p. 34).

As this passage suggests, Vernadsky was a man of his time who associated its aspirations with what appeared to be limitless capabilities of scientific and technological progress. However, this concerns another aspect of Vernadsky's heritage, his well-known teaching on the noosphere that will be addressed in Chapter 13.

Vernadsky's ideas were far ahead of his time and may have remained in obscurity for many years if it had not been for the new rapidly evolving contemporary branch of science known as ecology. This field, created mainly by biologists, concentrated the attention of scientists on the structure and properties of functioning biological complexes. However, it was only toward the second half of the 20th century that Vernadsky's concepts and bioecology came together in any significant way. The first to introduce the term "ecology" was the famous German naturalist and philosopher Ernst Haeckel (1834–1919). A disciple of Darwin, Haeckel used this term to indicate a branch of biology that studied the relationship between organisms and their environment. Defined by Haeckel as the "body of knowledge concerning the economy of nature" (in Stauffer, 1960) the term ecology was hardly used in scientific circles until the early 1900s. A particularly significant contribution to the development of this field was made by hydrobiologists, which can be easily explained by the fact that their object of inquiry was aquatic organisms. Such life forms are virtually impossible to study in isolation from their physical environment.

One of the pioneers in this area was the German zoologist Karl Möbius (1825– 1908). Based on an investigation of mollusk reproduction on the oyster banks of the North Sea, Möbius originated the notion of an interconnected community of organisms inhabiting a given homogeneous area on the sea floor. Coining the term biocenosis (Möbius, 1977) in order to describe this phenomenon, Möbius noted the way evolution determines the rigid adaptation of various species not only to one another—but also to the specific conditions of the local abiotic environment (biotope). Subsequently, this concept was applied to communities on land and in fresh water (e.g., pond biocenosis, lake biocenosis, or birch forest biocenosis).

Research in the area of organization at the supraorganismal level did not become truly widespread until the early 20th century. Contributions were made by biologists from many fields: botany, zoology, hydrobiology, forestry science, etc. Of particular importance in this connection was establishing certain general patterns typical of the way that various organism complexes (communities, biocenoses) developed in the course of their interaction with the environment. An example would be the vegetation succession process which has regular stages of development among different ecosystem types. Ecology is indebted for this discovery to two American botanists. The first was Henry C. Cowles (1869–1939) who studied vegetation on the banks of Lake

Michigan, a body of water that had been growing shallower over a very long period of time, thereby moving farther and farther away from its original shoreline.

Cowles correctly assumed that the age of a given community has to increase in proportion to movement away from the water's edge. This allowed him to reconstruct the detailed structure of this entire process. The youngest, most recently formed dunes are occupied by perennial grasses whose roots reinforce unfixed sands. These grasses are then replaced by more tenacious gramineous plants, that are followed by bushes and shrubs. Under this layer, trees begin to grow out of older, more reinforced dunes according to a fixed sequence: first pines, which are replaced by oaks and maples within a generation. Finally, at the greatest distance from the shore, beech trees, the most shade-loving ones in this climate, make their appearance (Odum, 1971).

Subsequently, the concept of vegetation succession was worked out in great detail by Cowles' disciple, the plant ecologist Frederic Clements (1874–1945). In his classic work entitled *Plant Succession* (1916), Clements studied the vegetation community as a one whole organism undergoing specific development: from youth to maturity. He demonstrated the adaptive capacity of biocenoses and their capability to evolve in the course of environmental changes. While various communities in the same area can be very different from one another in the early stages of their existence, they grow to be increasingly similar as time goes by. In the end, it turns out that only one mature or so-called climax community is typical of every area within given climate and specific soil conditions (Phillips, 1954).

In 1927 the English zoologist Charles Sutherland Elton (1900–1991) published *Animal Ecology*, a book that set the standard in population ecology by redirecting the attention of zoologists from single organisms to a given population as an independent whole. At this level, argued Elton, one could establish specific characteristics of ecological adaption and regulation. Having taken part in two Arctic expeditions shortly before the publication of this work, the author noticed cyclical fluctuations in the numbers of small rodents with a periodicity of 3–4 years. Elton processed long-term data on fur stocks in North America and concluded that hares and lynxes were also subject to cyclical fluctuations although their respective population numbers peaked approximately once every 10 years. In what is today considered a classic piece of research, he was the first to describe the structure and distribution of animal communities, introducing the ecological niche concept and formulating the *pyramid of numbers* rule. The latter is the consistent decrease in the numbers of organisms that accompanies movement from lower trophic levels to higher ones along the food chain, from plants to herbivores to predators, etc. (Elton, 1946).

Meanwhile, in the 1920s and 1930s ecology was enriched by precise methods of investigation thanks to the work of American Alfred J. Lotka (1880–1949) and Italian Vito Volterra (1860–1940). Lotka's *Elements of Physical Biology* (1925) was the first attempt to transform biology into a strictly quantitative science. This book developed mathematical models and calculations of inter-species interaction (e.g., a model describing the interlinked dynamics of prey and predator numbers) as well as biogeochemical cycles. Although Lotka was not yet using the term "ecology" proper, his attempts to apply physical laws to biological objects are a vivid illustration of a trend toward the expansion of research horizons undertaken through the ecological revolution. A year after the publication of Lotka's book, Volterra developed a mathematical model for the manner in which two species compete for one resource and demonstrated the impossibility of their sustained long-term co-existence (1926).

The theoretical inquiries of Lotka and Volterra attracted the attention of a young Soviet biologist called Georgii Gause (1910–1986) who proposed his own "more biological" modification of the equations describing inter-species competition processes. He tested these models experimentally with the help of laboratory bacterial and protozoan cultures, which made it possible to demonstrate that the coexistence of species was possible only when it is determined by differing environmental factors. In other words, different species occupy different ecological niches, which means that competition by more than one species for the same niche always produces a loser. This is known as the competitive exclusion principle. Gause's work was published in the U.S.A. in 1934 under the title *The Struggle for Existence* (in Russia it did not appear until 70 years later), in many ways facilitating the emergence of population ecology as a science. The emphasis he placed on the importance of trophic connections as the main pathway for energy flow through natural communities made an important contribution to the formation of the ecosystem concept.

However, the actual term ecosystem was introduced in 1935 by the English botanist Arthur George Tansley (1871–1955).[1] Tansley's greatest achievement was the successful integration of biocenosis with the biotope at the level of a new functional unit: the ecosystem. Other, more established sciences, such as physics, chemistry, and cytology, had long relied on their own basic units: the atom, the molecule, and the cell, respectively. For ecology this role was fulfilled by the ecosystem that can be defined as a temporally and spatially limited unified natural complex. The latter is constituted by living organisms and their habitat where living and non-living components are interconnected through metabolic activity and the distribution of energy flow.

A few years later and independently of Tansley, the Russian biologist V. N. Sukachev (1880–1967) used forest communities to develop the concept of biogeocenosis (1940). This was analogous to the ecosystem, being characterized by spatial limitations, as well as the homogeneity of natural and climatic conditions. On land, biogeocenosis can refer to a small area, such as a riverine meadow, or a tree and the soil underneath it corresponding to the sweep of the tree crown. Natural and climatic conditions include biotic as well as abiotic components of the environment unified by a common substance cycle and energy flow. Therefore, in territorial and hierarchic

---

[1] Although he had quite authoritative predecessors, it is common to cite 1935 as the birth of general ecology as an independent science. Among these precursors are Edward Ashael Birge (1851–1950) and Chancey Juday (1871–1944) who, in the very beginning of the 20th century, had studied lake communities in order to determine the role of organisms in the substance cycle and the transformation of energy. Equally noteworthy are the Germans Reinhard Demoll (1882–1960) and August Friedrich Thienemann (1882–1960) who formulated in the 1920s such ecologically important notions as biomass and biological production.

terms, biogeocenoses can be viewed as compartments or "cells" of the biosphere which itself is an ecosystem of the highest hierarchic level (i.e., the ecosystem of the planet; cf. Bernhard and Reimers, 1991).

The emergence of the ecosystem as a concept constituted a radical change in ecology as a whole. Up until that point, ecologists had been too disparate in terms of their research, but now a new broad thrust of inquiry was created. As before, a crucial role was played here by hydrobiologists whose object of study (aquatic organisms often inhabiting enclosed water bodies, such as ponds and lakes) was particularly characterized by the intertwining of chemical, physical, and biological processes. Thus, the American limnologist Edward Birge (1859–1950), who studied "lake respiration", used strict quantitative methods to establish the seasonal dynamics of water-dissolved oxygen content determined not only by water mixing and the diffusion of air-borne oxygen, but also by the activity of living oxygen-producer organisms (planktonic algae) and living oxygen-consumers (bacteria and animals).

Subsequently, in the 1930s these ideas were developed further in the work of Russian limnologists L. L. Rossolimo (1894–1977), G. G. Vinberg (1905–1987), and others. The last in this series of milestones was the elaboration of the so-called balanced energetic approach that made it possible to deal with the substance cycle and energy transformation in an ecosystem through the use of strictly quantitative indicators. This method required unity among biochemical processes taking place in various organisms (e.g., photosynthesis in all planktonic algae within a pond or in all plants within a forest), and on that basis one attempted to add up the results of their activity according to the amount of resulting organic substance and released free oxygen. In this manner it became possible not only to conduct a quantitative assessment of biological production in an aqueous or sylvan ecosystem, but also to work out mathematical models for these ecosystems' fundamentals.

Similar measurements were carried out in the United States under the direction of George Evelyn Hutchinson (1903–1991) who became famous not only through his own research (his *Treatise on Limnology* published in 1957 is still the most complete account of life in a lake)—but also thanks to his active support of young scientific talent. It is likely no accident that his scientific school had so much impact on the development of ecology in many countries of the world. Among Hutchinson's pupils, first place is assigned to prematurely deceased Raymond Laurel Lindeman (1915–1942). It would not be an exaggeration to say that Lindeman's short article entitled "The trophic-dynamic aspect of ecology" (1942) was a breakthrough in the field and is still referred to by ecologists all over the world. In this study the author worked out the general scheme of energy transformation in an ecosystem and outlined the key methods for the calculation of its energetic equilibrium. In particular, Lindeman demonstrated theoretically that during the transition of energy from one trophic level to another (from plants to herbivores to predators) the amount of available energy decreases. Thus, organisms in each successive level have access to only a small part (no more than 10%) of the energy at the disposal of living things occupying the level just below (Bolshakov *et al.*, 1996).

From this point on, ecosystem research evolved into one of the main directions in ecology, and the quantitative analysis of ecosystem components and functions constituted one of the key methods for predicting and modeling biological processes. And so, step by step, through the efforts of hundreds of scientists, ecology was creating new structures and occupying the freshly constructed floors of a building whose contours had been drawn in the work of Vernadsky. However, this was still short of grasping the biosphere as a global ecosystem. Vernadsky died the year that World War II ended, and his ideas were still underestimated by his contemporaries. Even the work that summed up his endeavors, *The Chemical Structure of the Earth's Biosphere and Its Environment*, was not published until 15 years after Vernadsky's death. Several decades had to pass before the notion of the biosphere as an entire integrated system began to take root in the minds of scientists. A prime example of such a thinker was the remarkable Russian biologist N. V. Timofeiev-Resovsky.

Having made his name in the pre–World War II decade while working in Germany on radiation genetics, Timofeiev-Resovsky spent the last years of his life dealing with questions of global ecology and in many ways anticipated the subsequent conception of numerous still "embryonic" problems in this field. As was the case for many other Soviet intellectuals, Timofeiev-Resovsky was arrested under Stalin's regime and sentenced to hard labor in a Siberian camp. After his release, he settled in the city of Obninsk since the path to big centers, like Moscow or Leningrad, was unavailable to a former inmate of the infamous Gulag. Thus, in 1968 at a meeting of the Geographical Society's Obninsk chapter, he presented a paper entitled "Biosfera i chelovechestvo" ("The biosphere and humanity"). Here is how his main thesis was formulated:

> "The biosphere is a gigantic living factory that transforms energy and substances on the planet's surface. It shapes the balanced composition of the atmosphere, as well as the composition of solutions in natural waters. Through the atmosphere, the biosphere determines the energetics of the planet and influences our climate. Let us recall the tremendous role played in the Earth's moisture cycle by the evaporation of water by vegetation and the planet's plant cover. Therefore, the Earth's biosphere shapes the entire environment in which humanity finds itself. Its neglect and any activity which undermines the proper functioning of the biosphere will not only compromise human food resources, as well as a whole range of industrial raw materials that we require, but will also harm the gaseous and aqueous environment of our civilization. In the final analysis, human existence on Earth is simply impossible without the biosphere or even within a deficient biosphere" (Timofeiev-Resovsky, 1996, pp. 59–60).

This paper was published as an eponymous article in a collection of scientific articles of the Geographical Society's Obninsk chapter. However, given the narrow distribution of this rather obscure publication, very few people read it and still fewer (perhaps a handful) were in a position to truly give Timofeiev-Resovsky's ideas their due. And so, as has often been the case with many Russian pioneers, the paper and the article went virtually unnoticed, among other reasons because the U.S.S.R.'s Academy of Sciences was unwilling to pay attention to a "disgraced" scientist. Timofeiev-Resovsky's contemporaries had missed the point that in developing

Vernadsky's teachings, he was one of the first to express a key notion about a full-scale regulation of the biosphere by life itself.

Admittedly, just as in Vernadsky's case, Timofeiev-Resovsky's thinking was still representative of a time when discourse about the "conquest of nature" appeared as the norm. Hence, he placed his hopes in the possibility of transforming the "gigantic living factory" (the biosphere) through human agency. This reliance on the prevailing ideology of the period placed the scientist in contradiction with some of his own ideas. However, in light of Timofeiev-Resovsky's achievement, this is a fairly insignificant failing. What matters more in this instance is that he was active on the "wrong side" of the Iron Curtain, which, as in the case of many other Soviet scientists, made his work virtually inaccessible in the West.

The advantage conferred through working on the "right side" of the Iron Curtain is illustrated by the English researcher James Lovelock (b. 1919) whose famous biospheric Gaia Hypothesis was met with great interest by the scientific community in the 1970s. Lovelock had worked as an engineer for NASA where he developed devices intended to search for life on other planets, a task associated with upcoming unmanned flights to Mars and Venus. In his student years, Lovelock had created a unique gas spectrophotometer for measuring minute gas concentrations in the atmosphere. Subsequently, this very device made it possible to determine the accumulation in the atmosphere of chlorofluorocarbons (CFCs) that were destroying the Earth's ozone layer. This activity led Lovelock to the idea that the existence of life on a given planet could be determined on the basis of atmospheric composition since the atmosphere is highly unstable and sensitive to all biogeochemical changes in the environment. Lovelock supposed that the atmosphere of "living" planets had to be characterized by a thermodynamic disequilibrium maintained through the activity of life. In the case of "non-living" planets, on the other hand, atmospheric composition is in equilibrium with the average chemical composition of the celestial body. All these ideas inspired the further elaboration of Lovelock's Gaia Hypothesis which was published in the form of a short article (1972) and then developed in a series of books and monographs (cf. Lovelock, 1979).

The image of Gaia (after the Ancient Greek Earth goddess) appears when one casts a mental glance at our planet from space. The Earth becomes a multilevel, multilayered, living organized whole. This can be expressed in terms of movement from a macro-level to a micro-level: biosphere–biocenosis–organisms–organs–cells. Here is what Lovelock wrote about the Earth as a whole:

> "The climate, the composition of the rocks, the air and the oceans, are not just given by geology; they are also the consequences of the presence of life. Through the ceaseless activity of living organisms, conditions on the planet have been kept favorable for life for the past 3.8 billion years. Any species that adversely affects the environment, making it less favorable for its progeny, will ultimately be cast out, just as will those members of a species who will fail to pass the fitness test" (Lovelock, 2000, p. 25).

Thus, Gaia is seen as a kind of self-organizing system or "super-organism" with self-regulating "geophysiological" properties, capable of maintaining homeostatically the parameters of the planetary environment at a level favorable to life. In this connection, the evolution of the biota is so closely linked with the evolution of its physical environment that together, they form a kind of single, self-evolving system whose properties partially resemble the physiology of a living organism (Kazansky, n.d.).

In his theories, Lovelock paid special attention to the Earth's bacterial community whose importance in the evolution of the biosphere from the beginnings of life to our time cannot be overstated. For approximately two billion years bacteria were the only life forms on the planet; as catalysts of biogeochemical cycles, they were responsible for shaping the biosphere. To this day they remain the foundation of the Earth's biogeochemical mechanism. Initially, the ancient community of prokaryotic microorganisms dominated the globe, covering its surface as a thin film of living mats, a kind of unified fabric.[2] While at first they monopolized the biogeosphere, as evolution progressed, the autocatalytic units of this community "migrated" and became integrated into more complex organisms in the form of specialized organelles (mitochondria and chloroplasts) within eukaryotic cells.

Therefore, the regulation of Gaia's physiological processes (reduction and oxidation, the combination of oxygen with carbon, etc.) is carried out today both by the direct descendants of unicellular prokaryotes, such as soil bacteria, and by their descendants in eukaryotic cells, such as mitochondria (oxidizers) and chloroplasts (reducers). This catalytic hypercycle in a way unifies the tiniest living organisms with the planetary macrosystem in terms of maintaining the climatic and biogeochemical parameters of the environment (Eigen and Schuster, 1979). This started over 1.5 billion years ago when the modern oxygenic atmosphere was formed (Kazansky, n.d.).

It is easy to discern the obvious similarities between the Gaia Hypothesis and our modern understanding of Vernadsky's biosphere concept. Incidentally, because there were no adequate English translations of *The Biosphere* and because, as Lovelock himself admitted, English-speaking authors tend to be "deaf" to those writing in other languages, the originator of the Gaia Hypothesis did not learn about Vernadsky's work until the 1980s (i.e., in a way he "rediscovered" the biosphere half a century after Vernadsky). However, there are differences between the two approaches. First, Gaia is not really the biosphere but rather the Earth as a whole. Lovelock draws a parallel between Gaia and a cross-section of an old tree trunk where the living part (biosphere) is but a thin layer of cambium under the bark while most of the mass is non-living wood substance, a product of cambial activity over many years. Another difference is Lovelock's skeptical attitude toward the belief in humankind's ability to "conquer" nature and compel it to do our bidding, which characterized Vernadsky's time.

The question is whether the Gaia Hypothesis (the term "hypothesis" comes from the author himself) is scientific in the strict sense of the word, beyond the daring ideas and the philosophical or moral subtext. It ought to be noted that some of Lovelock's

---

[2] In microbiology such mats are benthic (dwelling at the bottom of water bodies, and in or on the ground) microbial communities and biofilm characterized by dense and layered biomass packaging with a vertical gradient of chemical and physical factors.

"geophysiological" hypotheses have been substantiated experimentally. For example, in 1981 the author expressed the supposition that the global climate may have the capacity to stabilize through the self-regulation of the CO_2 cycle attributable to the biogenic increase of rock weathering (Lovelock and Watson, 1982). In geophysiological terms, CO_2 is a key metabolic Gaia gas that influences not only the climate but also plant production and the generation of atmospheric oxygen. Volcanic activity is a source of CO_2 in the biosphere. When dissolved in rainwater, CO_2 interacts with volcanic rocks rich in carbonates and silicates (chemical weathering).

Indeed, the results of work conducted by David W. Schwartzman and Tyler Volk published in *Nature* confirmed the fact that microorganisms together with plants are capable of accelerating rock weathering by factors of tens or hundreds (1989).[3] Plants transfer CO_2 from the air into the soil, thereby changing the local concentration of this gas by a factor of 10–40, while the bulk of dead vegetation undergoes bacterial oxidation and also turns into CO_2 at points of contact with calcium compounds, silicates, and water. The products of weathering are washed out by river flow into oceans where they are used by algae for skeleton building. Furthermore, oceanic algae take in CO_2 directly from the air and, upon dying, form chalky sedimentary deposits. In this manner, the biota, by affecting the concentration of atmospheric CO_2 (one of the greenhouse gases), participates in the regulation of average global temperature.

One can cite other examples of proven cyclically closed causative chains that typify the geophysiology of the Gaia Hypothesis.[4] However, the picture becomes murkier when it comes to Lovelock's central idea: that Gaia is a global intracorrelated superorganism. In its time, this notion encountered vigorous criticism from numerous famous evolutionary biologists (e.g., Doolittle, 1981; Dawkins, 1982). The problem is that biospheric evolution in the framework of the Gaia Hypothesis is interpreted as the biosphere's individual development (epigenesis) and the improvement of its self-regulatory properties. However, from the standpoint of traditional science, such highly complex and strictly intracorrelated systems as Gaia must inevitably deteriorate over time and fall apart. Living organisms, in contrast to nonliving matter, are also characterized by a level of organizational complexity without equal. However, in order to maintain this complexity and order, nature uses a unique mechanism of competitive interaction on the part of separate individuals. This causes the elimination of those who have lost this internal order and, as a result, the ability to compete (Gorshkov and Mararieva, 2002). It is in this manner that the evolutionary process leads to the reproduction and regeneration of the unique complexity characterizing living matter.

---

[3] Vernadsky noted the role of biogenic weathering back in the 1930s.

[4] In 1973 it was possible to confirm one of Lovelock's predictions, namely, the emission of dimethyl sulfide and methyl iodide (products of vital activity on the part of dying plankton) from the ocean to dry land. This turned out to be the only known source of sulfur and iodine transfer from ocean to land. When sulfuric compounds reach land, they facilitate the growth of plants that accelerate rock leaching. The biogens formed as a result of this process are washed out into the sea and in their turn facilitate the growth of algae production.

Now, all this having been said, Gaia is a single entity, which means that it cannot reproduce. As Richard Dawkins notes, natural population selection is impossible in the case of the best-adapted planet. Therefore, we cannot discuss the more or less long-term maintenance of Gaia's self-regulatory capability unless we assume the intervention of an intelligent creator's ordering will. Or, as Ford Doolittle ironically observed, a biological species committee would have to meet on a yearly basis in order to agree on the following year's climate and chemical composition of the planet. Lovelock could not counter this criticism, which led to discrediting of the idea that life uses its own means to form a favorable environment for itself.

In what follows, we will outline the approach by which the St. Petersburg biophysicist Professor V. G. Gorshkov, and author of a theory on the biotic regulation and stabilization of the environment, tried to solve this problem in the 1980s. For now, however, let us return to the already cited work of N. V. Timofeiev-Resovsky who preceded Lovelock with a solution to the problem in question. Timofeiev-Resovsky noted the structural unit of the biosphere that serves as the framework for natural population selection: not individuals or species—but biological communities. These are biogeocenoses or "elementary units of the biological cycle constituting the biogeochemical work that takes place in the biosphere." Here is how Timofeiev-Resovsky went on to explain this phenomenon:

> "Most biogeocenoses are in a state of relatively long-term dynamic equilibrium and constitute considerably complex self-regulating systems. Therefore, of particular importance is the problem of studying the causes, mechanisms and conditions related to maintaining such dynamic equilibrium in biogeocenoses. [Without knowing these mechanisms] one cannot conceptualize correctly the real progression of natural evolutionary processes which always unfold in dynamic biocenoses and their larger complexes, namely, landscapes" (Timofeiev-Resovsky, 1996, p. 63).

It is easy to see the extent to which such a structured system of "biospheric cells" differs from the Gaia Hypothesis. If the work required to maintain the biogeochemical cycle is not carried out by the biota in general or by a sort of superorganism that keeps the biota together, but rather by separate biotic communities and their populations, then there is room for their competitive interaction. It is this mechanism, whereby poorly functioning "cells" are displaced and replaced, that protects the biosphere from degradation and disintegration, thereby indefinitely maintaining its ability to sustain global biogeochemical equilibrium. This is also called *stabilizing selection*. However, the same can be said about individuals that are intercorrelated within every species or the species themselves which determine the "look" of a biogeocenotic community and with respect to which biogeocenosis really does appear as a kind of superorganism (a term first introduced by F. Clements in 1916).

Table 9.1 summarizes the contributions of various scientists to the formation of a systematic biosphere concept along the two conceptual lines of thought introduced by Haeckel and by Vernadsky.

In what follows, we will be dealing in more detail with stabilizing selection, in connection with the biotic regulation and stabilization of the environment.

Table 9.1. Contributions of various scientists to the formation of a systematic biosphere concept along the two conceptual lines of thought introduced by Haeckel and by Vernadsky.

Haeckel line	*Vernadsky line*
Ernst Haeckel (1834–1919)—German evolutionary scientist, follower of C. Darwin. Introduced the concept of ecology as a biological field for the study of relationships of organisms with their environment.	Eduard Suess (1831—1914)—Austrian geologist who was the first to use the term "biosphere" to indicate one of the Earth's shells, along with the atmosphere, hydrosphere, and lithosphere.
Karl Möbius (1825–1908)—German zoologist and hydrobiologist. Using an oyster bank on the shoals of the North Sea, he developed and grounded the notion of an intracorrelated community of organisms inhabiting a given homogeneous area of sea floor that he called biocenosis (1877).	Vladimir Vernadsky (1863–1945)—Russian mineralogist and geochemist. Proposed and developed the concept of the biosphere as an integral and intracorrelated world of living matter connected by a system of biogeochemical cycles with abiotic spheres: namely, the atmosphere, hydrosphere, and lithosphere. Vernadsky demonstrated that the chemical state of the Earth's external crust is entirely under the influence of life and governed by living organisms.
Henry Cowles (1869–1939)—American botanist. While studying the vegetation on the shore of Lake Michigan, he discovered a pattern of stages in the development of various ecosystem types. He was the first to describe the process of biological succession.	Nikolai Timofeiev-Resovsky (1900–1981)— Russian biologist and one of the founders of molecular genetics and radiobiology. He devoted the last years of his life to global biological issues. In his publications and presentations of this period, he compared the biosphere to a gigantic living factory that forms the earthly environment. In 1968 he was the first to point out the role of biogeocenoses as elementary units in the biological cycle. He also pointed out the possibility of competition among biogeocenoses, which forms the basis for stabilizing evolutionary selection.
Frederic Clements (1874–1945)—American botanist who worked out in detail the concept of succession (1916). He introduced the notion of a mature climax community as a concluding stage of biological succession. He demonstrated the adaptive nature of biocenoses: their capacity to evolve in response to environmental changes.	James Lovelock (born 1919)—English scientist with a background in electrical engineering who invented the electron capture detector. He was also an environmentalist and a futurist. While working for NASA, he developed equipment for detecting life on other planets in connection to upcoming unmanned travel to Mars and Venus. This led him to propose an original concept according to

Haeckel line	*Vernadsky line*
	(Lovelock *cont.*) which the Earth is an integral super-organism where the evolution of life is closely linked to changes in its physicochemical environment. He named this notion after the Greek goddess of the Earth Gaia. The Gaia Hypothesis has made it possible to reconsider the global processes of the substance cycle. Many of Lovelock's theoretical predictions were confirmed experimentally. However, he was unable to explain how Gaia—which is a highly intracorrelated system—could avoid inevitable degradation and disintegration for hundreds of millions of years. This caused his ideas to be criticized by evolutionary scientists.
Charles Elton (1900—1991)—English zoologist who laid the foundations of population ecology. He introduced the notion of ecological niches and formulated the rule of ecological pyramids: the decrease in the number of organisms at each shift from lower to higher trophic levels.	Victor Gorshkov (born 1935)—Russian theoretical physicist who has devoted his efforts to the study of the general principles behind the functioning of life. On this basis he developed the concept of the biotic regulation of the environment. As opposed to Lovelock, Gorshkov connected the maintenance of environmental parameters suitable for life with the life activity of competing independent biotic communities (biogeocenoses). As it reacts to environmental disruptions by changing the synthesis or destruction of organic substances, biota can fix the excess of certain biogens in the environment or, conversely, make up for their shortfall. This allows for the regulation of biogen concentrations at a level suitable for life. In his work Gorshkov has been able to demonstrate that the modern global ecological crisis has to do first of all with the destruction of natural ecosystems on huge areas of land. His argument is that the transition to sustainable development is possible only if a significant portion of these destroyed ecosystems is restored. A key *(continued)*

Table 9.1 (*cont.*).   Contributions of various scientists to the formation of a systematic biosphere concept along the two conceptual lines of thought introduced by Haeckel and by Vernadsky.

Haeckel line	*Vernadsky line*
	(Gorshkov *cont.*) accomplishment stemming from the concept of the biotic regulation of the environment is the law of distribution of energy streams in the biota. According to Gorshkov's calculations, more than 90% of the energy flow in natural, undisrupted biota has to do with plant organisms, bacteria, and fungi that carry out most of the work for the stabilization of the environment. And, less than 1% of the energy flow is accounted for by large animals, including man. In this manner, one can assess the limits of the energy corridor beyond which human civilization must not venture if we are concerned about the stability of the environment.

Haeckel line (cont.)
Alfred Lotka (1880–1949)—American mathematician and author of a book entitled *Elements of Physical Biology* where we witness the first attempt to transform biology into a strictly quantitative science. He developed mathematical models of interaction between species (e.g., the conjugated dynamics of predator and prey numbers) and biogeochemical cycles.
Georgii Gause (1910–1986)—Russian biologist who modeled processes of interspecies competition on the basis of laboratory bacterial cultures and protozoans. He formulated the law of competitive exclusion according to which two species cannot occupy the same ecological niche since one of them will inevitably displace the other. He stressed the importance of trophic links as a fundamental path of energy streams through natural communities. This made an important contribution to the formation of the ecosystem concept.
Arthur Tansley (1871–1955)—English botanist who in 1935 introduced the concept of ecosystem. This was the birth of general ecology as an independent science. Tansley's main accomplishment was the successful attempt to integrate biocenosis with the biotope at the level of a new functional unit—the ecosystem. This unit became what the atom was for physics, the molecule for chemistry, and the cell for cytology.
Vladimir Sukachiov (1880–1967)—Russian biologist and forestry scientist. Using the example of forest communities, he developed the concept of biogeocenosis in 1940. This was analogous to the ecosystem and was characterized by the limited spread and homogeneity of natural climate conditions. This notion included both biotic and abiotic environmental components. In terms of territory and hierarchy, biogeocenosis can be viewed as a minimal unit or cell of the biosphere unified by the general substance cycle and energy flow.

Haeckel line (cont.)
Raymond Lindeman (1915–1942)—American ecologist who in 1942 developed the general scheme of energy transformation in an ecosystem and the main methods for calculating its energy balance. He was able to prove theoretically that when energy is transferred from one trophic level to another, its amount is reduced every time by an order of magnitude. In this manner, organisms at every subsequent level can access no more than 10% of the energy available to organisms at the previous level.
Eugene Odum (1913–2002)—American biologist and one of the founders of ecology as an independent scientific discipline. He became famous for his pioneering work on ecosystem ecology, as well as his textbooks: *Fundamentals of Ecology* (with Howard Odum, 1953), *Ecology* (1963), etc. These books played an important role in the formation of ecology as an independent course of scientific study. Odum collected what had existed in the form of disparate journal articles and monographs and formulated this material anew in the form of fundamental questions of ecological science. He thereby made ecology accessible to wide circles of biologists and helped introduce ecological disciplines into the university curriculum.
Robert MacArthur (1930–1972)—American ecologist who laid the foundation of the mathematical study of populations and developed a series of forecasting models for ecosystems. Together with E. R. Pianka, he studied the problem of how an organism makes the optimal choice of feeding locations in a heterogeneous environment. He also developed the concept of allowable overlapping mutual niches among competing species (the optimal use of a patchy environment). In his book entitled *The Theory of Island Biogeography* (with E. O. Wilson), he established the general principles of the formation of island communities. He was significantly instrumental in the transformation of ecology as a predominantly descriptive science into an experimental field as well.
David Tilman (born in 1949)—American ecologist who was an important representative of a new direction in the field. This movement was interested not so much in any given resulting structures, but rather in the processes that determine these structures (i.e., physiological and behavioral mechanisms). Therefore, in contrast to the hypothetical–deductive approach, this became known as the mechanistic approach. Tilman's most notable works had to do with the study of limiting resources on the basis of diatom algae and cereals. He was able to prove that ecologically proximate species can coexist only if they are limited by various resources (e.g., different threshold concentrations of biogens in water).

References

Bernhard, J. M.; C. E. Reimers (1991). "Benthic foraminiferal population fluctuations related to anoxia: Santa Barbara Basin," *Biogeochemistry* **15**, 127–149.

Bolshakov, V. N.; S. V. Krinitsin; F. V. Kriazhimsky; Kh. P. Martines Rika (1996). "The problems of modern social perceptions of key ecological scientific concepts. *Ekologia* **3**, 165–170. Available at *http://biospace. nw. ru/evoeco/lit/bolschakov. htm* [in Russian].

Dawkins, R. (1982). *The Extended Phenotype*. Oxford, U.K.: W. H. Freeman & Co.

Doolittle, W. F. (1981). "Is nature really motherly?" *CoEvolution Quarterly* **29**, 58–63.

Eigen, M.; P. Schuster (1979). *The Hypercycle: A Principle of Natural Self-Organization*. Berlin: Springer-Verlag.

Elton, C. S. (1946). *The Ecology of Animals*. London: Methuen.

Gause, G. F. (1934). *The Struggle for Existence*. Baltimore, MD: Williams & Wilkins.

Gorshkov, V. G.; A. M. Makarieva (2002). "Greenhouse effect dependence on atmospheric concentrations of greenhouse dubstances and the nature of climate stability on Earth," *Atmos. Chem. Phys. Discuss.* **2**, 289–337. Available at *http://www. atmos-chem-phys.org/acpd/2/289/*

Kazansky, A. B. (n.d.) *The Gaia Phenomenon of James Lovelock*. Available at *http://www.evol. nw.ru/labs/lab38/kazansky/fenomen.htm* [in Russian].

Lindeman, R. L. (1942). "The trophic-dynamic aspect of ecology," *Ecology* **23**, 399–418.

Lovelock, J. E. (1972) "Gaia as seen through the atmosphere," *Atmospheric Environment* **6**, 579–580.

Lovelock, J. E. (, 1979). *Gaia: A New Look at Life on Earth*. Oxford, U.K.: Oxford University Press.

Lovelock, J. (2000). *Gaia: The Practical Science of Planetary Medicine*. Gaia Books.

Lovelock, J. E.; A. J. Watson (1982). "The regulation of carbon dioxide and climate: Gaia or geochemistry," *Journal of Planetary Science* **30**(8), 795–802.

Odum, E. P. (1971). *Fundamentals of Ecology*, Third Edition. Philadelphia: W. B. Saunders.

Phillips, J. (1954). "A tribute to Frederic E. Clements and his concepts in ecology," *Ecology* **35**(2), April, 114–115.

Schwartzman, D. W.; T. Volk (1989). "Biotic enhancement of weathering and the habitability of Earth," *Nature* **340**, 457–460.

Stauffer, R. C. (1960). "Ecology in the long manuscript version of Darwin's 'Origin of Species' and Linnaeus' 'Oeconomy of Nature,' *Proceedings of the American Philosophical Society* **104**(2), 235–241.

Timofeiev-Resovsky, N. V. (1996). "The biosphere and humanity." In: A. N. Tiuriukanov, V. M. Fiodorov, N. V. Timofeiev-Resovsky (eds.), *Thoughts on the Biosphere*. Moscow: RAEN [in Russian].

Tiurukanov, A. N.; V. M. Fiodorov; N. V. Timofeiev-Resovsky (1996). *Thoughts on the Biosphere*. Moscow: RAEN [in Russian].

Timofeiev-Resovsky, N. W.; Zimmer K. G.; Delbruk M. (1935). "Über die Natur der Genmutation und Genstructur," *Nachrichten von der Wissenschaften zu Göttingen, Neu Folge* **1**(13) [in German].

Vernadsky, V. I. (1987). *The Chemical Structure of the Earth's Biosphere and Its Environment*. Moscow: Nauka [in Russian].

Vernadsky, V. I. (1988). *The Philosophic Ideas of a Naturalist*. Moscow: Nauka [in Russian].

Vernadsky, V. I. (1998). *The Biosphere: Complete Annotated Edition*, translated by D. Langmuir. New York: Springer-Verlag/Copernicus Books.

Volterra, V. (1926). "Variazioni e fluttuazioni del numero d'individui in specie animali conviventi," *Mem. R. Accad. Naz. dei Lincei, Ser. VI* **2**.

Zavarzin, G. A. (2004). *Lectures on Microbiology in the Context of Natural History*. Moscow: Nauka [in Russian].

10

The constancy of the planetary environment in light of the biotic regulation mechanism

10.1 ABIOTIC FACTORS IN THE FORMATION OF THE EARTH'S CLIMATE

The role of the hydrosphere in climate stabilization—The flipside of the greenhouse effect—Glaciers as cold accumulators—The causes of the great glaciations—"The climate marble" balancing on the tip of an unstable equilibrium—The Earth's biota as a stability factor in the Earth's climate

As opposed to James Lovelock, Victor Gorshkov did not work for NASA and was not connected with the Soviet space program; however, the *View from Space* played an important role in the formation of his theory. In other words, Gorshkov tried to project onto the Earth the physical conditions characterizing the celestial bodies adjacent to us. Gorshkov's idea was stimulated by a question that had preoccupied him for a long time, namely, the astonishing and puzzling stability of biota over billions of years. Indeed, how was it possible that over this truly "cosmic" time period, given all the geologic and climatic perturbations experienced by our planet, the seemingly vulnerable relay of life—or as L. Itelson puts it, "the unbelievably weak tiny flame flickering in the furious winds of the universe"—was never interrupted? (Itelson, 2000, p. 109). What made biota victorious through thick or thin: asteroid collisions with the Earth, climate cool-downs and glaciations, monumental volcanic eruptions, continental drift through plate tectonics, and variations in World Ocean levels? And in general, Gorshkov wondered what ensured life's place within the thin layer separating living things from the molten bowels of the Earth, on the one hand, and, on the other, the eternal cold and deadly cosmic radiation of interplanetary space. Was it a combination of coincidental and non-coincidental factors (i.e., the extraordinarily fortunate positioning of the Earth's orbit at a precise spot with respect to the Sun) and against this background the appearance and stabilization of life according to a predictable mechanism?

Indeed, life on a planet is possible within a relatively narrow temperature range where water exists in a liquid state. Only a few degrees below the freezing mark constitutes an extreme temperature for the vast majority of species. A limited number of warm-blooded animals are able to survive actively and for a long time in below-zero conditions (emperor penguins can even multiply under such temperatures). And only a few thermophilic bacteria species can exist at temperatures above 60°C.[1] The optimal temperature range for the majority of the Earth's life forms is 5°C–25°C.[2] As radioisotope studies of rock and sedimentary deposits demonstrate, the average surface temperature on our planet has remained within this range for the past 600 million years. It went down to +5°C during glacial periods and up to +25°C at peak warming points. The modern average global temperature is +15°C and has stayed at this level for many centuries with variations not exceeding several tenths of a degree.

And so, how does one explain this astonishing stability of the temperature and climate on Earth? Some of the causes are obvious and well known to modern science. First of all, there is the stable nature of solar radiation and the light energy reaching the Earth from the Sun—of the order of 1,360 J/m^2. Of that amount around 30% is reflected back by our cloud cover, atmospheric dust, and glaciated or snow-covered areas of the Earth's surface, producing no thermal effect. Another 23% is dissipated in the atmosphere and is used up in the process of water evaporation. The combination of the 47% that directly reaches the surface of the planet, plus the 23% dissipated in the atmosphere, heats up the surface of the ocean and land, which then in turn leads to warming of the atmosphere through secondary long-wavelength infrared radiation.

No less significant factors include the position of our planet's circumsolar orbit, as well as the approximately 66.5° axial tilt angle of the Earth's rotation with respect to its orbital plane. This facilitates the warming of higher latitudes within the framework of seasonal change. Finally, a huge role in the maintenance of the Earth's climate and temperature is played by the planet's powerful atmosphere and hydrosphere, the latter being capable of accumulating a huge amount of heat energy. The mass of the Earth (6×10^{23} t) makes it possible to retain such a mighty layer of air and water. In order to grasp the tremendous amount of heat accumulated by the World Ocean, we need to recall that the latter occupies 70.8% of the planet's surface and contains 1,320 million km^3 to 1,380 million km^3 of water whose specific thermal capacity is 1 cal/g per 1°C or 4,186.8 J/kg per 1°C (the highest thermal capacity of all existing liquids). As a result, the World Ocean's heat reserves are 21 times greater than the yearly amount of heat energy received by the Earth from the Sun. Therefore the hydrosphere can be justifiably viewed as a key temperature stabilizer on the planet's surface (Marfenin, 2006).

---

[1] Extreme thermophilic bacteria have been discovered living at the bottom of the ocean in hot sulfide-containing springs. They can exist under high pressure and even in temperatures as hot as +115°C. [However, it seems possible that if climatic excursions had been greater, life may have evolved in such a manner as to survive under these harsh conditions.]

[2] It seems likely that had climatic excursions been greater, life would probably have evolved in such a manner to tolerate these harsh conditions.

The warming of the Earth does not occur evenly; however, thanks to airmass displacement in the lower strata of the atmosphere and powerful ocean currents (from the equator to the poles and back), this unevenness is leveled out to such an extent that life can exist at virtually every latitude. Furthermore, the atmosphere is constantly taking in moisture that evaporates from the surface of the World Ocean and amounts to around $500,000 \, \text{km}^3$ or 86% of all evaporated water. Through the evaporation process a great deal of energy is transferred to the atmosphere and then transported by air currents (i.e., winds, cyclones, and hurricanes are generated by the global energy redistribution machine). In the same manner, before precipitating in the form of rain or snow, water vapor can be transported by air currents for thousands of kilometers. Around 90% of water evaporating from the surface of the World Ocean falls back into it, constituting the so-called big or oceanic hydrologic cycle. The remaining 10% of precipitation falls on land and constitutes the lesser or continental hydrologic cycle. Furthermore, the main mass of water precipitation returns along with river runoff to the World Ocean while a small part of it is retained by glaciers.

Depending on altitude, geographic latitude, and season, the concentration of water vapor in the atmosphere varies from 0.5% to 4.0% for an average of 1% (*Climate Change*, 2007). Together with CO_2 (its current atmospheric concentration is 0.04%), these vapors play a very important role in the greenhouse effect, retaining part of the infrared heat radiation emitted by the atmosphere and the surface of the Earth (dry land and water bodies) heated up by sunlight. The total concentration of all the other greenhouse gases (methane, nitrous oxide, etc.) does not exceed 3×10^{-4}%. It has been estimated that the greenhouse effect accounts for an approximately $40°C$ increase in average surface temperature. In other words, the presence of small to minute amounts of greenhouse gases in the atmosphere protects the Earth from freezing and thereby prevents the planet from sharing the fate of Mars.

However, it is easy to see that the greenhouse effect has a dark side, namely, an increase in atmospheric greenhouse gas concentration can bring about a corresponding rise of the Earth's surface temperature, although in the case of CO_2, the main absorption band is saturated so further increases in concentration produce diminishing increases in temperature. Given the high complexity of the climate system, this can have extremely unpredictable consequences. For example, as the temperature of the World Ocean's surface layer goes up, the solubility of CO_2 in water decreases. The disruption of the buffering equilibrium between the ocean and the atmosphere causes the latter to experience an increase in the concentration of CO_2 (i.e., intensification of the greenhouse effect). On the other hand, as glaciers melt and permafrost regions are destroyed, they release CO_2 and methane, which is accompanied by the accumulation of water vapor in the atmosphere. The result is something like a chain reaction as the climate keeps warming up more and more. This is why the experts, extremely alarmed by the threat of climatic imbalance, are demanding the adoption of urgent measures for the limitation of human impact on the environment.

While the World Ocean is the planet's heat accumulator, the glaciers can be justifiably deemed cold accumulators whose contribution to the formation of the Earth's climate and the stabilization of surface water runoff from dry land areas cannot be underestimated. Thus, for example, the glaciers contain 69% or about 2/3

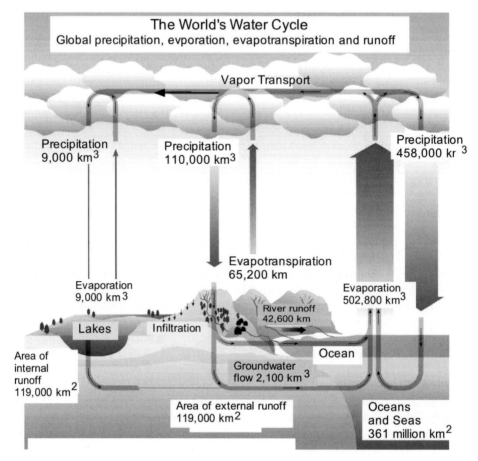

Figure 10.1a. Water is transported in different forms within the hydrological cycle or "water cycle". The widths of the arrows are proportional to the flow rates (UNEP, 2002).

of the world's freshwater reserves while their combined runoff into the oceans $(3,450 \, km^3$ of water) constitutes about 8% of the total runoff from all surface water sources and approximately 3% of precipitation over dry land. Acting as natural water reservoirs, the glaciers facilitate the redistribution of runoff from atmospheric precipitation during any given year. This is particularly important for regions where rivers are fed to a significant extent by glaciers: for example, the Yukon (23% glacier runoff), the Kuban (6%), the Indus (8%), the Syrdaria (6.5%), the Amudaria (15%), etc. In Central Asia, where only 5% of the territory is occupied by glaciers, the proportion of glacier runoff in the summer months reaches 50%.

At the same time, glaciers contribute to the cooling of the atmosphere, which is determined by the increase in Earth's albedo,[3] on the one hand, and, on the other

---

[3] Albedo is the reflective power of the Earth. The term refers to the fraction of incident radiation that is reflected back into space.

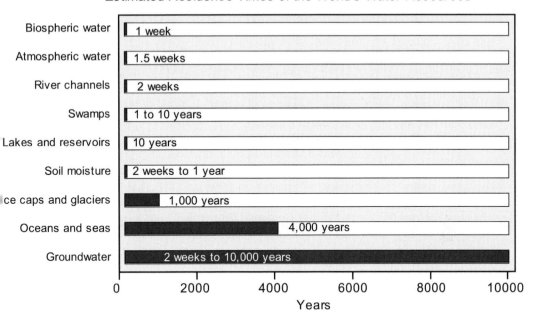

Figure 10.1b. Estimated residence time of the world's water resources (UNEP, 2002).

hand, by heat expenditure for the melting of ice. Occupying 3% of the planet's surface, the glaciers account for 5% of solar radiation reflected back into space, which increases the Earth's average yearly albedo from 0.29 to 0.30 and leads to the cooling of the surface layer of air by approximately 1°C. The combined effect of reflected sunlight from glaciers and melting of glacial ice produces a cooling effect of approximately 2°C (Govorushko, 2006). Of course, the cooling effect due to melting only takes place during periods of global warming. During periods of global cooling, the opposite is true.

Therefore, if the World Ocean has always facilitated the conservation of heat on the planet, the glaciers have had the opposite function, and the relationship of these forces has differed significantly from one geological epoch to the next. Science knows numerous episodes in the history of the Earth when the average surface temperature dropped by 5° to 6° or more as glaciers covered huge areas of dry land. The thickness of the ice sheets pressing down on the continents could reach 4 km while ocean levels (as compared with the present) dropped by 100 m to 120 m. The last such glaciation ended about 12,000 years ago and the entire Quaternary geological period (which started 1.8 Myr to 1.6 Myr ago and includes the modern Holocene epoch) has even been called "glacial" because of repeated major glaciations.

The reasons for these periodic glaciations are still not entirely understood. Most likely they are the result of several combined factors of varying degrees of importance while others act to catalyze trends, once started, via positive feedback. The prevalent belief is that ice ages and interglacial periods are induced by changes in the Earth's orbit (obliquity, eccentricity, and precession of seasons) that change the input of solar energy to high northern latitudes on timescales of many tens of thousands of years, and these changes are amplified by positive feedback effects such as the increased reflectivity of spreading ice and snow. However, this theory also has difficulties, and has been challenged by a number of scientists. The role of the oceans in delivering heat to higher latitudes may also be significant. Along with the slow changes from glacial to interglacial conditions over tens of thousands of years, there have also been wild oscillations in the Earth's climate when it underwent large temperature excursions within decades or centuries.

And so, despite our powerful atmosphere and oceanic layer, the Earth's climate system is in a state of rather unstable equilibrium that can easily be disrupted under the influence of various internal and external factors. On the basis of what we know about almost 4 billion years of life on Earth, there are no objective reasons for assuming that the Earth cannot shift into a stable state that is unsuitable for life. Examples of such sterile equilibrium can be observed among our neighbors in the Solar System. Hence, Mars has a glaciated surface with an average temperature of $-40°C$. Venus is the other extreme: a greenhouse pressure cooker where the temperature goes up to $+475°C$. While Mars is farther from the Sun, and Venus is closer to the Sun than Earth, nevertheless it is physically possible for the Earth to undergo climate extremes. Geological evidence suggests that the Earth has passed through "Hothouse Earth" and "Ice Ball" phases in its distant past (Rapp, 2009).

This is all the more so because, according to existing models, the initial stages of planetary evolution were more or less similar on Mars, Venus, and Earth. The three planets are of comparable size, weight, and average density. The study of Venus with the help of unmanned space probes Venus-9 and Venus-10 revealed the presence of cragged rock protrusions close in composition and appearance to basaltic rock. The latter is widespread all over the Earth and appeared around the same time as its Venusian counterpart. We should also mention that the total CO_2 content is the same on both planets and, according to calculations made by the Russian geochemist A. P. Vinogradov, equals around 2.1×10^{17} t on Earth or Venus. The difference is that on Earth most of the CO_2 is bound in various areas of the lithosphere or dissolved in the depths of the ocean. On Venus, where a hydrosphere cannot exist, almost all the CO_2 is concentrated in the planet's atmosphere, 97% of which consists of carbon dioxide. This is why Venus is so immensely hot (Vinogradov, 1959; Vinogradov et al., 1970).

However, perhaps one of the factors determining the maintenance of a life-sustaining temperature regime is the very biota inhabiting the Earth. By taking part in the formation of biogeochemical cycles, biota facilitates the creation of conditions favorable to itself. An example would be the binding of excess CO_2 and its consequent elimination from the atmosphere by the Earth's biota or, conversely, the release of CO_2 in the process of decomposition of organic matter that compensates for the insufficiency of the gas in the atmosphere. In order to justify this hypothesis, Professor V. Gorshkov attempted to calculate the balance of the Earth's climate on the basis of Aleksei Liapunov's well-known sustainability principle.[4] Here the Earth's biota was left out of the equation. Without dwelling on the mathematical aspects of these calculations, we can only say that given the graphic structure of the curve determined by Gorshkov, today's climatic and temperature parameters of the Earth appear to be located at the highest point of the curve (i.e., a zone of high instability). According to this model, minor variations in average global temperature could inevitably lead to irreversible climate change. The climate might then be stabilized only under conditions approaching those on Mars or Venus (Gorshkov *et al.*, 2000).

In popular science literature the instability of the temperature–climate equilibrium on Earth is often compared to a marble at the top of a pointed pyramid. For a certain amount of time the marble can keep its balance in this precarious position; however, even an insignificant external force can cause the marble to roll down and end up at the base of the pyramid. The latter location is where the marble is truly stable or physically selected. However, what if one were to cut out a cavity at the top of the pyramid—corresponding to the marble's shape? This would provide for a new physically selected state in which the marble could exist for a long and indefinite period. Something like that, according to V. Gorshkov, takes place in the case of the Earth's climate where the above-mentioned "cavity of stability" at the top of the temperature curve is created for the climate by biota itself, ensuring no longer physically, but rather biotically selected stability (see Figures 10.2 and 10.3).

Indeed, geological evidence suggests that at certain times in the almost 4-billion-year history of life on Earth, the average surface temperature came under the influence of processes external to the biosphere (stemming from inside the Earth or space) and rose up to 50°C—or even 90°C according to some estimates ("Hot-house Earth", Kasting, 1993). The opposite has also been the case when the entire Earth may have been sheathed in a layer of ice and snow and remained that way for long periods ("Ice Ball Earth", Hoffman and Schrag, 2002). Life may have survived through such epochs at localized volcanic vents. At a lesser level, ice ages have spread thick sheets of ice over large northern areas, spreading as far down as New York State. So the depth of the biotically selected "climatic cavity" has probably varied from epoch to epoch as the stability of the planet's climate underwent very serious trials many times. However, since the "climatic marble" still managed to maintain its position within parameters amenable to life, the explanation for this must be sought first and foremost in the properties of biota itself. Therefore, the existence of the latter appears to be not only the consequence but also the cause of the temperature and climate conditions favorable to life.

In this case Gorshkov had to deal with the question of the mechanism behind this stabilizing impact of biota on the environment. What were the characteristic traits of

---

[4] Liapunov, Aleksei Andreievich (1911–1973): Russian mathematician, associate member of the Soviet Academy of Sciences, and recipient of the Computer Pioneer Medal (1996) by the IEEE (Institute of Electrical and Electronics Engineers) as the founder of Soviet cybernetics and programming.

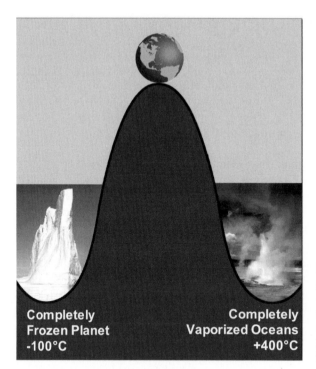

Figure 10.2. (Left) Complete glaciation of the planet −100°C. (Right) Complete evaporation of the oceans +400°C. The Earth is precariously balanced between these extremes.

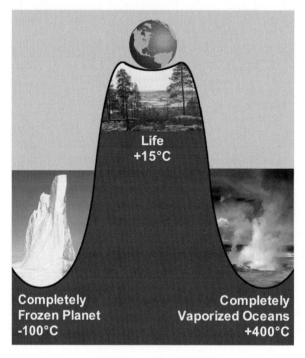

Figure 10.3. (Left) Complete glaciation of the planet −100°C. (Right) Complete evaporation of the oceans +400°C. Life exists in between.

all living things in light of this impact? Years of research yielded Gorshkov's *Concept of the Biotic Regulation of the Environment* which made it possible to use a large amount of theoretically based and empirical data in order to make a connection between three elements: (a) understanding of the global biospheric processes, (b) the function and place of the Earth's biota within these processes, and (c) the role of the destructive potential underlying human economic activity.

Anyone interested in the mathematical basis used by Gorshkov for developing his theory should read the monograph: *Biotic Regulation of the Environment: Key Issues of Global Change* or visit the author's site at *http://www.bioticregulation.ru/ index.php*. As for our part, we would like to outline in the most accessible way the key ideas underlying this concept, illustrating them with examples of the means by which biota organizes its abiotic environment. We also intend to present the most important conclusions stemming from this work that have a direct bearing on the problem of sustainable development. In Section 10.3 we will deal with Gorshkov's research on energy flow patterns in biota and the biosphere's carrying capacity interpreted in this context.

10.2 A FEW EXAMPLES OF THE BIOTA'S DIRECTED INFLUENCE ON THE ENVIRONMENT

Finite biospheric carbon resources and the "waste-free technologies" of the modern biota—The elimination of carbon from the atmosphere in preceding geologic epochs— The role of oceanic biota in the carbon cycle—The soil as a reservoir of organic carbon and moisture accumulator—The role of biota in soil formation—How plant biota regulates the continental moisture cycle—The biotic forest pump as a key condition for life on dry land—Why Europe has not become a desert.

As far back as V. I. Vernadsky's time, scientists understood the exceptional nature of the role played by the Earth's "living substance" in shaping the global substance cycle thanks to the high speed of chemical reactions involving life. Today it appears obvious, for example, that life has changed the face of the Earth, and that it created the planet's oxygen-rich atmosphere some 2 billion years ago. However, the quantitative characteristics of certain biogen cycles—including the carbon cycle that constitutes the basis of organic molecules and therefore the basis of life itself— have been studied only relatively recently. This, in turn, made it possible to make significant strides towards grasping many processes occurring in the biosphere. Thus, it was determined that today, as opposed to the situation in ancient geological epochs, the reserves of inorganic carbon accessible to the biosphere are significantly limited, constituting approximately 1,500 Gt C ($1,500 \times 10^9$ t). This is primarily CO_2 dissolved in the World Ocean, as well as its small admixture in the atmosphere— approximately 0.04%. However, most of the CO_2 dissolved in the oceans is located at depths below 200 m and is hardly used by phytoplankton. Furthermore, there is relative little upward flow of CO_2 stemming from the abysmal oceanic strata as a result of vertical water exchange. Therefore, biota has to be extremely sparing when it comes to the use of inorganic carbon.

However, life strategy looked very different in the case of ancient biotic communities billions of years ago when the atmosphere was far richer in CO_2 and may have consisted primarily of N_2, CO_2, and water vapor. This conclusion can be derived from the huge oil and coal deposits from those epochs—energy sources used widely in our time. According to data on remnants of organic substances unprocessed by living organisms and accumulated in ancient geological formations, we can assume that in the initial stages of biospheric evolution the waste material produced by life was simply removed from the environment and buried in sedimentary deposits (petroleum). Or, as was the case in the Devonian and the Silurian, the accumulation of rotting plant remains was more rapid than the rate of their bacterial decomposition (coal). This type of biotic strategy can be deemed highly entropic (i.e., it is accompanied by the significant mortification of organic substances and considerable generation of dead matter; Krasilov, 1992).

Therefore, the need for more effective use of environmental trophic resources exerted evolutionary pressure for the creation of "waste-free technologies" so to speak—the basis of the manner in which modern natural ecosystems function. This was facilitated in the first place by the increasing complexity characterizing the structure of biotic communities through the diversification of life forms and corresponding ecological niches. Another important factor was the degree of inter-correlation of species within communities that ensured the multiple utilization of organic substances created by plant producer organisms in the process of photosynthesis. This is the origin of the multi-tiered system typifying a more or less enclosed substance cycle and including several sequential stages involved in the processing of organic production (trophic chains)—starting with autotrophic producers at the system's entrance and ending with decomposer organisms, such as the fungi and bacteria, at its exit. The latter life forms process non-decomposed organic matter down to the final low-molecular-weight compounds, thereby making it accessible for assimilation by plant root systems. This ensures the possibility of reutilizing biogenic chemical elements required for life (i.e., that which is dead ends up given back to the living).

It has been estimated that the Earth's producer organisms create more than 100 Gt per year of organic matter (equivalent to pure carbon mass) part of which is used for maintaining their own life—respiration, growth, and reproduction—while the rest is utilized by consumer organisms of the first, second, and the following orders: herbivores, carnivores, and detritivores. From this it is evident that if the above biochemical cycle were not enclosed, all of the carbon reserves available to the biosphere would be depleted within a very short time.

However, modern biota, having adapted to an oxygenic atmosphere, can tolerate neither carbon excess nor its insufficiency. In this connection we should recall the dangerous consequences of the excessive greenhouse effect. It ought to be mentioned that the main mass of inorganic carbon is concentrated not in the atmosphere or the World Ocean, but rather in the bowels of the Earth. The current ratio of its content between the lithosphere, hydrosphere, and atmosphere (28,570:57:1; Marfenin, 2006) is under constant threat. Indeed, because of volcanism, great quantities of inorganic carbon are constantly entering the atmosphere and hydrosphere through magma degassing and from reef faults in the ocean. In volcanic gases, CO_2 constitutes ~15%, water vapor 75% to 80%, whereas all the other gases (CO, CH_4, NH_3, H_2S, etc.) aggregate to less than 10%. At the same time, it has been determined that in the Earth's remote history there were periods of far greater volcanism as compared to today, and yet CO_2 concentration in the atmosphere—even given its significant fluctuations—maintained its order of magnitude for hundreds of millions of years (Broecker *et al.*, 1985; Barnola *et al.*, 1991).

Today we understand the fate of the carbon emitted into the atmosphere. In past geological epochs it was all safely tucked away in sedimentary rock: that is, the fossilized remains of biota from former biospheres (chalk, limestone) and in hydro-carbon minerals (petroleum, coal, peat, oil shale, etc.). All this is related to biota's activity for maintaining environmental homeostasis—a mechanism that removed "excess" carbon from circulation for thousands of years. The magnitude of this process was evaluated on the basis of minute impregnations of organic carbon—kerogen—that is found in geological deposits. It turns out that the rate of its accumulation over the past 600 million years has been around 0.01 Gt C per year (Budyko *et al.*, 1985). Such is the functioning of this global mechanism for the stabilization of the carbon cycle, that it ensures a correlation between the influx of inorganic carbon from the bowels of the Earth and its deposition in sedimentary rock. Some scientists suppose that an analogous mechanism was at work in more remote epochs when the atmosphere was more strongly carbonic. For example, the Russian geologist academician V. E. Khain and his American colleague K. J. Hsu think that ancient cyanobacteria in proterozoic seas also facilitated the removal of CO_2 from the Earth's primary atmosphere, burying the gas in the depths of stromatolitic carbonates (Khain, 2007; Hsu, 1992).[5]

A special place in the regulation of the global carbon cycle belongs to the biota of the World Ocean. There is 57 times more CO_2 dissolved in seawater than is contained in the atmosphere—40,000 Gt C as compared with 700 Gt C (Bolin, 1983). This dissolved CO_2 does not diffuse into the Earth's atmosphere because the surface water layer in the ocean is poor in CO_2. This 100 m to 200 m thick upper layer is called photic because it is penetrated by light from the visible solar spectrum, which creates

---

[5] As was recently established (Rosing *et al.*, 2006), ancient microorganisms at the bottom of proterozoic seas processed the upper layer of the basalt crust which went down to the depths of the Earth together with bottom sediments (also processed by microorganisms). As a result of remelting this mass produced granitoids—the foundation of the continental crust. To quote V. E. Khain (who refers to Rosing *et al.*), "because they are not as dense as the basalts and gabbroids of the oceanic crust, blocks of continental crust floated up, as it were, above the asthenospheric mantle and manifested themselves mainly as dry land or shallow seas" (Khain, 2007). In other words, the ancient biota played a key role in the formation of continents whose appearance (around 3.5 billion years ago) coincides approximately with the appearance of life.

conditions necessary for photosynthesis. It is warmer than the deeper ocean layers where practically no warming takes place. The photic layer floats, as it were, over the colder regions from which it is separated by a zone of abrupt change in temperature and density known as the thermocline. Furthermore, the photic layer is thoroughly mixed by wind action, which, in combination with the chemical properties of carbonates dissolved in water, ensures relatively quick absorption of atmospheric CO_2 and the equalization of its concentration in air and water according to the Henry–Dalton Law. However, the most important property of the photic layer is that it provides a sanctuary for phytoplankton that generates the main mass of aqueous organic matter through photosynthesis. This aqueous organic matter feeds all the consumers of the ocean. This process is accompanied by the fixation (binding) of CO_2 that accumulates in the ocean at the rate of $\sim$120 Gt C per year (Green *et al.*, 1984, Vol. 1).

Therefore, with respect to the global carbon cycle, the photic layer acts as a kind of buffer that, on the one hand, absorbs the CO_2 accumulating in the atmosphere and, on the other hand, prevents excess CO_2 dissolved in seawater from entering the atmosphere. The latter mechanism is known as the biotic pump because the CO_2 taken out of the ocean depths as a result of storms and upwelling[6] becomes part of the organic synthesis process at the ocean surface. Subsequently, as living organisms that inhabit this environment die off, this bound carbon sinks back to the ocean bottom where it accumulates in the form of dissolved organic matter. Alternatively, the carbon may be incorporated into the calcareous skeletons of organisms in the form of calcium carbonate ($CaCO_3$) and end up buried in the midst of sediments, thereafter being released in the process of decomposition. To quote N. V. Koronovsky,

"The role of biogenic sediment accumulation used to be underestimated until quite recently. Today it has been determined that 50–65% of all sedimentary mass corresponds to biogenic matter with yearly accumulations reaching approximately 350 billion tons (dry substance equivalent). The matter dissolved in oceanic waters is assimilated by life forms that filter these waters. Only half a year is required for living organisms to filter all the water of the World Ocean" (Koronovsky, 2002).

In this manner the sea biota contributes to maintaining the composition of oceanic water that it requires. This composition has remained essentially unchanged for the entire duration of the Phanerozoic (approximately the past 600 million years). Similarly, this mechanism facilitates the elimination of excess atmospheric carbon

---

[6] Upwelling is the rising of oceanic water from the thermocline and deeper layers under the influence of wind action which forces warm water off the surface. Other factors at play include water density differences, the configuration of the coastline, and a number of others. This process is particularly significant in connection to the transportation from the ocean depths of various nutrients that enrich the upper layers and increase their bioproductivity.

by binding it through organic synthesis processes and then partially burying it in sediments at the ocean bottom.[7]

If we could travel a billion years back in time and look at the planet's dry landmass from a bird's eye view, we would see neither any signs of vegetation nor any "landscape" as we understand the term today. Instead of rivers flowing through their beds, we would be looking at a kind of boundless delta with innumerable arms and tributaries. And instead of a clear coastline separating dry land from the sea, we would behold many kilometers of partly flooded terrain (neither land nor sea), etc. As K. Iu. Es'kov put it, "there are serious reasons for assuming that dry landscapes of the modern type simply did not exist back then".[8] Or, according to A. G. Ponomarenko, "the existence of real continental water reservoirs (flowing or stagnant) appears quite problematic before vascular vegetation could somewhat slow down the rate of erosion and stabilize the coastline" (1993, cited in Es'kov).[9] In other words, living organisms did not merely emerge onto dry land—they formed it in a manner of speaking, and higher (vascular) plants made a fundamental contribution to the shaping of modern landscapes. This embodied the biblical verse "let the dry land appear" (Genesis 1:9) by making soil.

The question is how soil could be developed on this lifeless ground washed out by rains and subject to unregulated sheetwash. As the work of the Russian scientist V. V. Dokuchaiev (1846–1903) showed, soil is a very complex mixture of organic minerals that appeared as a result of physicochemical forces acting on biota's parent rock. One such force is the destruction of rock through erosion—a process which until recently was thought to take place under the influence of sunlight, wind, temperature fluctuations, and freezing and thawing cycles of water in crevasses and cracks. Only in the past few decades has it been possible to establish the tremendous role of living organisms in this process—in the first place bacteria and fungi which accelerate erosion by a factor of 100 to 300.

Once they "land" on rock undergoing erosion, these organisms dissolve and destroy its surface layer where, as they die off, their dry dead remains form and fill cavities, hollows, and furrows. Furthermore, biota colonizing rock surfaces extracts from them the chemical elements it needs to sustain itself and brings about the formation of rather aggressive organic acids which significantly accelerate the dissolution and hydrolysis of minerals. In this manner, biota becomes an active supplier of loose material that accumulates and becomes capable of retaining moisture along with the organic and inorganic compounds dissolved in it. This creates a suitable substrate for the germination of plant seeds. The resulting rootlets branch out and penetrate the soil layer, which facilitates the structuring of the soil and its reinforcement (i.e., the formation of landscapes that we know today).

---

[7] We are not dealing here with the details of the nitrogen, phosphorus, and sulfur cycle, and that of other so-called minor biogens whose proportion in the general mass of organic matter is not great (compared with that of carbon) amounting altogether to no more than 1%.

[8] *http://www.paleo.ru/paleonet/publications/eskov/08.html*

[9] On the other hand, some believe that a very long time ago land was entirely devoid of moisture with all precipitation falling over ocean (Gorshkov and Makarieva, 2007).

The foregoing should make clear the great extent of biota's role in the process of soil formation, despite the relatively small proportion of organic substances in the soil (around 10%). However, this modest amount involves the entire composition of the soil in a permanent cycle, thereby bestowing upon the soil the properties of both living and non-living substances. Plant roots discharge the products of their life activity into the soil; these products are then partly involved in chemical reactions and partly consumed by fungi, as well as by other microorganisms. The latter process plant and animal remains in order to extract the substances required for their life activity, thereby contributing to the decomposition of organic matter down to its simpler molecular compounds that then become available once again for assimilation by the roots of plants.

Another contribution to the recycling of non-decomposed organic matter is made by various invertebrates, such as insect larvae, myriapods, and earthworms that feed on fallen leaves and pass them through their intestinal tracts. It has been estimated that over a period of 100 years, virtually the entire 0.5 m thick soil cover of dry land in the temperate latitudes passes through the gut of earthworms. These animals grind up and mix the mineral and organic elements in the soil, thereby improving its structure. Furthermore, earthworms dig passageways that facilitate soil aeration and root growth (Lapo, 1987; Sorokhtin and Ushakov, 1991; Green et al., 1984, Vol. 2).

Given its ability to accumulate organic and mineral substances, as well as moisture, the soil acts as a reservoir and source of life with respect to the biota. The humus accumulates colossal reserves of carbon and biogenic elements; however, the organic substances involved are not the ones contained in plants and animals but rather newly formed ones. These are first and foremost humic acids containing up to 50% to 60% carbon—the reason for the characteristically black color of chernozem soil. Finally, the highly dispersive, porous structure of the soil gives its constituent particles a large aggregate surface area. This provides a moisture reservoir ensuring the retention of much rain and melt water. Ancient dry land was devoid of these properties and returned precipitation in its entirety to the ocean (unregulated sheet-wash). And so, having emerged onto land millions of years ago, biota "oceanized" it by creating the soil as a kind of ocean fed by alluvium. The other evident creative act was the shaping of the modern freshwater system consisting of swamps, rivers, and lakes.

And now let us turn to the contribution made by the plant biota to the regulation of the continental moisture cycle. As has been stated above, the soil, given its porous structure, is an effective mechanism for moisture retention on land. However, this retention is short-lived since land is elevated above sea level (i.e., according to the laws of physics, soil moisture must flow downward from higher levels, collecting into rivulets, streams, small and big rivers until it finally reaches the ocean). According to recent data, the volume of this yearly runoff is around $43,000\,\text{km}^3$. It has been calculated that within a period of four years, all the land-based water accumulated in lakes, swamps, and mountain glaciers would end up in the ocean if it were not replenished by atmospheric precipitation. About 2/3 of this precipitation forms right above land from the evaporation of terrestrial moisture sources while approximately

1/3—or to be precise 35/100 of yearly precipitation (assuming distribution over the entire land surface)—corresponds to oceanic evaporation. In other words, if evaporated moisture from the ocean did not fall in the form of rain and snow onto land, the latter would become completely free of freshwater within fewer than roughly 10 years.

Without a doubt the process referred to as "nature's kitchen" (the displacement of atmospheric fronts, the generation of electrical storms, cyclone formation, etc.) constitutes a series of very complex phenomena which still require much study and resist smooth formal mathematical characterization and modeling. Nevertheless, in the past few years a rather promising attempt has been made to use known physical laws in order to connect the functioning of plant biota with the transfer of ocean moisture to land. This would demonstrate plant biota's stabilizing influence on this aspect of the global substance and energy cycle as well. This model deals with undisturbed forests; in fact, the authors—Prof. V. G. Gorshkov and Dr. A. M. Makarieva—refer to the mechanism of this moisture transfer as the "biotic pump of atmospheric moisture" (Gorshkov and Makarieva, 2007). In order to make sense of this mechanism, we need to take a closer look at the transpiration process, namely, the release of water from the leaf surface of plants in the course of photosynthesis. This is the key to understanding the phenomenon in question.

Transpiration—akin to blood circulation in animals—is the constant movement of water and substances dissolved in it within the plant. This water comes from the soil, traveling through the root system of plants and along the trunk xylem vessels (conductive plant tissue) toward the leaves. The motion from the roots to the leaves occurs along the so-called "water potential gradient". This gradient decreases with the increased concentration of salts and other substances in the water, which is facilitated by the selective permeability of cell membranes. The speed of water movement along trunk vessels is considerable, amounting to approximately 1 m/h in grassy plants and up to 8 m/h in tall trees. Xylem vessels are dead pipes with narrow openings of 0.01 mm to 0.2 mm in diameter. The ascent of water along such pipes to the top of a big tree requires pressure of the order of 4,000 kPa.[10] However, even when water is transported along the thinnest of vessels by capillary forces alone, it cannot rise higher than 3 m whereas the height of certain trees can reach 50 m and even 100 m—as in the case of the California *Sequoia* or the Australian *Eucalyptus*.

This phenomenon can be explained by cohesion theory. Thus, the ascent of water from the roots is determined by its evaporation from leaves, which leads to water depletion in leaf cells and consequently an increase in the concentration of substances dissolved in water. This results in the lowering of the water potential. Therefore, water from xylem fluid with higher water potential enters leaf cells through selectively permeable cell membranes. However, as water leaves xylem vessels, the water column experiences an increase in tension that is transferred down the stalk all the way to the roots. This tension is related to the molecular cohesion capacity of water determined by the polarity of water molecules or their dipole moment. The latter implies that under the influence of electrostatic forces, water molecules are attracted to each

---

[10] 1 kPa (kilopascal) = 0.01 atm.

other—become "glued" as it were—and are simultaneously held together by hydrogen bonds. Cohesion causes tension in the xylem vessels to reach such great magnitudes that the entire column of water can be drawn upwards. Estimates of tensile strength for a column of xylem fluid vary between 3,000 kPa and 30,000 kPa. The recorded water potential in leaves is of the order of 4,000 kPa. It appears, therefore, that the sturdiness of a xylem fluid column is sufficient to withstand the created tension (Green et al., 1984, Vol. 2).[11]

Finally, at the very last stage, water seeks to exit the plant because the water potential of the ambient, moderately moist air is several tens of thousands of kilo-pascals lower than in the plant itself. However, this departure takes place mainly in the form of water vapor, and this requires additional energy known as the "latent heat of evaporation". This energy is provided by sunlight that acts in the end as the force behind the transpiration process at all its stages—from the soil to the roots to the stalks to the leaves. Clearly, water is essential for ensuring the life-sustaining activity of a plant, which includes its photosynthetic needs. What may be less clear is the intensity of this process. After all, the plant itself uses on average less than 1% of the water it takes in from the soil whereas the remaining 99% returns directly to the atmosphere. In light of this, the level of transpiration—given sufficient illumination, moisture humidity, and ambient air temperature—can be very high. Thus, such herbaceous plants as cotton or sunflowers can lose up to 1 L to 2 L of water within a 24-hour period while a 100-year-old oak tree can lose more than 600 L (Green et al., 1984, Vol. 2). It is therefore appropriate that transpiration is sometimes referred to as a "necessary evil"; however, this is very naive.

Admittedly, most plants have evolved adaptation strategies that allow the regulation of this process and the retention of moisture if necessary. This includes the shedding of leaves during seasonal cold periods or drought, as well as water stockpiling in mucous cells and the cell walls of various plant parts. This also includes the presence of stomata that are special pores in the epidermis located in leaves and partially in green stalks. Gaseous interchange takes place through these openings out of which up to 90% of water evaporates. Thanks to special guard cells, the stomata can close during dry periods or at night when photosynthesis stops, which slows down the process of transpiration. Another adaptation for reducing transpiration, which forms under dry climatic conditions and moisture deficits, is the thickening of the cuticle—the waxy layer covering the epidermis of leaves and stalks. Nevertheless, under normal conditions and given a large leaf surface typical primarily of forest vegetation, water loss through transpiration can be very great. It can considerably exceed evaporation from water body surfaces equal in area to the tree crown's projection onto the ground. It is therefore not surprising that a natural forest

---

[11] The second force at work in the movement of water along the xylem is root pressure. This movement is determined by the active secretion into xylem fluid of salts and other water-soluble substances that lower the water potential of the fluid. As a result, the differential of osmotic pressure causes water to enter the xylem from neighboring root cells. However, this mechanism alone—creating hydrostatic pressure of the order of 100 kPa–200 kPa—is usually not sufficient to ensure the ascent of water all the way to the tree crown.

endowed with a high leaf area index (the ratio of the illuminated leaf area to the area of tree crown projection onto the ground) and given sufficient exposure to solar irradiance, can evaporate moisture much more intensively than the open ocean surface at the same temperature.

However, transpiration is not the only source of water vapor forming above the forest canopy, since trees are capable of using their crowns to accumulate or intercept a significant amount of atmospheric rainwater or snow. These two forms of moisture also make a sizable contribution (up to 30%) to the evaporation process generated by the forest. The latter circumstance is particularly important with respect to boreal coniferous forests where layers of snow wrapped around the trunks and crowns of trees ensure continuation of the flow of evaporation even in winter when transpiration is absent. In this manner, a forest that has not suffered a disruption is capable of evaporating moisture practically year-round and much more intensively than the open ocean surface at the same temperature, approaching (on average) the maximum possible magnitude of evaporation that is limited only by the flow of solar energy. Thus, calculations indicate that the maximal evaporation of water over the forest, corresponding to the flow of solar energy absorbed by the Earth's surface, equals ~2 m/yr, whereas the average global evaporation from the ocean surface is only ~1.2 m/yr (L'vovitch, 1979).[12] Furthermore, taking into account that the total leaf surface of plant biota is four times greater than the entire land surface of the planet, we can see that the aggregate transpiration of the forest can successfully compete with oceanic evaporation. From the position occupied by the authors of the forest moisture pump theory, this plays a key role in the formation of the continental moisture cycle.

The point is that the soil layer, as was pointed out earlier, cannot retain moisture for a long time without replenishment, losing it inevitably in the form of river runoff that in the end takes it to the ocean. Therefore, the problem of moisture retention on dry land is inextricably linked with compensating for these losses through reverse moisture movement from the oceans. However, the farther we move away from the coastline, the greater is the proportion of precipitation derived from the ocean that returns to the ocean through river runoff because of the continents' elevated position. At the same time, passive atmospheric geophysical currents that transport moisture from the ocean diminish as they progress deeper into the continent. This decrease in current strength is exponential. To be accurate, the above-mentioned tendency applies mainly to non-forested areas with low steppe-like vegetation bordering directly on the coastline. In such cases, for every 400 km of penetration into the steppe, savannah, or prairie, the flow of moisture and the intensity of precipitation undergo a diminution of approximately a factor of 2.

Gorshkov and Makarieva analyzed data on the gradient of precipitation over vast areas of land on five continents corresponding to the indicated criterion. Their findings demonstrate that the passive geophysical transfer of moisture toward bush-covered or grass-covered land devoid of a continuous tree canopy can provide normal conditions for life only in a zone within a few hundred kilometers of the

---

[12] This unit is actually cubic meters of water per square meter of area per year.

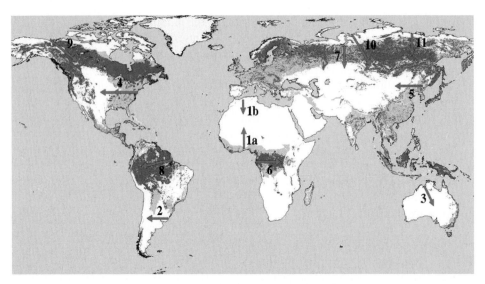

Figure 10.4. Geophysical regions studied as to the dependence of average yearly precipitation on distance from the ocean. Numbers on map indicate regions: 1a & 1b—West Africa, 5°E longitude; 2—South America, 31°S latitude; 3—Northern Australia; 4—North America, 40°N latitude; 5—East Asia, 42°N latitude; 6—Africa, Congo River Valley; 7—Ob River Valley; 8—Amazon River Valley; 9—Mackenzie River Valley; 10—Enisey River Valley; 11—Lena River Valley.

ocean. In the case of extensive continental areas, the average figure is about 600 km (Figures 10.4 and 10.5). That said, the water regime in such regions is determined by random fluctuations and seasonal changes in precipitation brought from the ocean, as is the case in monsoon areas with their abundant rains in the summer and very dry winters.

However, in view of these factors, how is one to explain the existence of well-watered areas deep inside continents—thousands of kilometers removed from the ocean (e.g., Siberia, the Amazon Basin, or Equatorial Africa)? It would probably be difficult to answer this question if we were to only consider passive geophysical currents, ignoring the active transfer of oceanic moisture based on the forest moisture pump. The essence of these processes can be summarized as follows. In a stationary atmosphere the weight of a gas at any given altitude is balanced by the weight of a gas column above this point. With increasing altitude, the atmospheric pressure drops. Anyone acquainted with mountain climbing knows this from personal experience: namely, it is more difficult to breathe in the highlands because the air there is thinner and contains correspondingly less oxygen.

As opposed to the other elements of the air, water vapor can exist in two phases—liquid (rain droplets and fog) and gaseous—and under certain circumstances water vapor can be transformed from gas to liquid. This phenomenon is known as condensation—exemplified by the formation of dew that settles on grass, bushes, or certain quickly chilling surfaces and objects in the cool of summer evenings. This

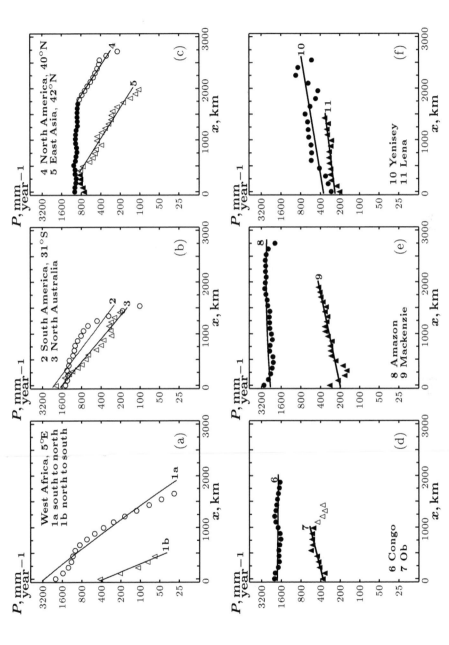

Figure 10.5. Dependence of precipitation amount P (mm/yr) on distance x (km) from the ocean in deforested regions (empty shapes) and forest-covered regions (filled shapes). Regions numbered as in Figure 10.4.

happens because a drop in temperature slows down the process of water evaporation, and the dynamic balance between its gaseous and liquid components shifts in the direction of the liquid. As they lose their kinetic energy, considerably fewer water molecules seek to leave the liquid phase, while the number of water vapor molecules returning to the gas phase remains unchanged—at first. During daytime, the concentration of water vapor in warm, dry air is likely to be below its possible maximum (the so-called level of saturation). However, when this air is cooled down at night, it may reach the saturation stage, and water vapor will begin to condense quickly and enter the liquid phase. In physics this critical temperature mark is known as the dew point, and it can be reached artificially by lowering the temperature of moist air or that of objects coming into contact with the air.

Something similar happens to water vapor with rising altitude. As is well known, with increasing altitude, air temperature drops by approximately 6°C/km. Thus, at an altitude of 10 km where most jets fly, the external temperature is almost 60°C below the surface temperature. For all other gas constituents of air, such a drop in temperature is not critical, but for water vapor the effect is significant. The concentration of water vapor in the air cannot rise above a certain maximum level called "saturated concentration". This maximum concentration decreases exponentially with decreasing temperature—an approximately two-fold decrease for every 10°C of temperature drop. If water vapor, like the other components of the air, were not a condensable gas, it would remain in gravitational equilibrium at any altitude so that its pressure would undergo a two-fold drop for every increase of 9 km in elevation. But in this case water vapor would be oversaturated in the entire atmospheric column, which is impossible. Therefore, in the real atmosphere, the relative surplus of water vapor continuously condenses and disappears from the gaseous phase. This is accompanied by a decrease in the weight of water vapor in an atmospheric column that can no longer compensate for its pressure in lower atmospheric strata. When condensation occurs, the atmospheric pressure of moist air falls. This brings about rising currents of moist air that contains residual water vapor. The latter reaches the upper atmospheric strata and condenses further, forming clouds and precipitation, such as rain and snow. Therefore, as the condensing water vapor leaves the air, this is constantly compensated for by water vapor brought by the rising currents of moist air from the ground level. These currents, to use Gorshkov's and Makarieva's image, are like the ancient Greek titan Atlas holding up the Earth's cloud cover.

And here we have reached the essence of the forest moisture pump. If the intensity of the rising currents is determined by the amount of water vapor condensation in the upper strata of the atmosphere, the power of these currents will increase with a rising intensity of evaporation from the Earth's surface that feeds them. Therefore, moist surface air will be sucked in from areas with less evaporation into areas with more intense evaporation. If, as has been shown above, evaporation above undisrupted forestland exceeds evaporation above the ocean surface, then ocean moisture will be pumped by the forest farther and farther into the depths of the continents, thereby compensating for river runoff and ensuring year-round soil moisture. However, this presupposes that the tree canopy extends all the way to the coastline, as is the case in the Congo and Amazon Basins or the northern rivers of

Russia and Canada. In the latter regions the taiga canopy is adjacent to the swampy areas of the far north that have access to the ocean. At the very least, woodland should be removed from the coastline at a distance below the attenuation distance of passive geophysical transfer.

It ought to be noted at this point that, given the difference in evaporation intensity, the process of pumping moisture from the cold ocean located at higher latitudes than the river basin itself appears to constitute a simpler task for biota than, for example, the transfer of moisture in the equatorial zone where this difference is much less significant. And yet, it is this difference that determines the speed of horizontal currents of pumped-in moist air, which would explain the existence of the great Siberian river basins covered by forests that stretch for many thousands of kilometers. As for the tropical latitudes, here the transfer inland of moisture from the ocean requires a much higher level of transpiration that would seem capable of preventing the transfer of atmospheric moisture from the forest to the ocean. In other words, continental moisture loss diminishes with increasing transpiration, which would explain the rising intensity of the latter noted by Brazilian ecologists during dry spells in the Amazon forests (da Rocha *et al.*, 2004)

However, all this applies only when the forestland is removed from shore at a distance less than the attenuation distance of passive geophysical transfer. The destruction of tree cover along the coastline within the attenuation distance ($\sim$600 km) interrupts the action of the biotic pump, so that the forest deep within a continent loses its capacity to compensate for river runoff. Soil moisture flows down into the ocean, the forests dry out, and river basins cease to exist. All these irreversible changes can happen within a very short period of time—of the order of 4–5 years required for the runoff of freshwater accumulated in mountain glaciers, lakes, and swamps.

It appears that something like this took place 50,000 to 100,000 years ago in Australia when the continent was first settled by humans. It is quite natural to suppose that the new arrivals acted in the usual manner, namely, they settled the coastline first, all the while destroying forests along the entire perimeter of the continent. When the attenuation distance of the deforested zone reached that of passive geophysical currents, this turned out to be, alas, sufficient—even in the absence of anthropogenic activity deep within the continent—to cut the biotic pump off from ocean moisture. The result was the demise of the Australian forests and the triumph of the Australian desert that today covers a huge part of the continent— 4 million km^2. Apparently the only way to account for the absence of any paleodata traces left by this transformation is the latter's brevity. Perhaps this is why most deserts border oceanic coastlines or have access to inland seas. The foregoing suggests that geography is closely related to the history of human settlement whereby the possession of new territories began from the coastline.

It may seem that Western Europe, which has practically lost its natural forests but has not yet undergone desertification, constitutes a case defying the above line of reasoning. However, this is the proverbial exception that proves the rule. Europe's felicitous lot is indebted to its unique geographic position. It is surrounded by inland seas, and its coastlines are close from any point. As a result, not a single West

European area is farther away from the coastline than the attenuation distance of the
geophysical transfer of sea moisture. The latter circumstance is apparently responsible for the illusion that the practice of deforestation can be exported to other regions
on the planet where such a policy is sure to be (or has already proven to be)
disastrous. At the same time, even in Western Europe we have observed a sharp
increase in the number of catastrophic flooding and dry periods, which is the result of
deforestation in mountainous regions and the general reduction of forested areas in
the first place with the consequent serious disruption of the natural hydrological
regime (the melting of mountain glaciers, etc.).

The desert environment may be viewed as practically closed off to moisture. This
is because the complete absence of transpiration, instead of causing moist air to be
sucked inland from the ocean, brings about the transfer of dry air to the ocean. On
the other hand, on steppe-like or savannah-like landscapes, as well as artificially
irrigated lands and pastures, the intensity of evaporation during the warm season
can be greater than the same process over the ocean's surface. In warmer periods the
oceans and seas are the source of a horizontal moist air current whose strength is a
negative function of the distance it covers (Figure 10.6). This current is known as
the summer monsoon or the rainy season. In winter or during colder periods,
evaporation over bushes and grass stands is less intense than its oceanic counterpart.
Therefore the moisture retained here is pulled from the land to the sea, creating the
dry winter monsoon (see Figure 10.6). In this manner, although the vegetation of
steppe-like ecosystems does ensure the maintenance of a certain moisture reserve and
the existence of evaporation currents on dry land, the absence of a continuous forest
canopy with a high leaf area index prevents such areas from increasing evaporation to
a level at which the current of moisture from the ocean to land can compensate

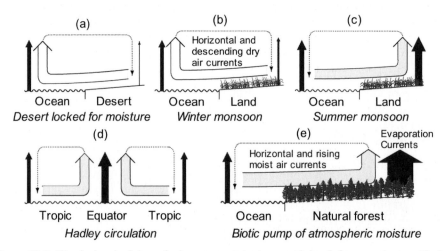

Figure 10.6. Physical principles of air movement at ground level from regions with less
evaporation to those with more evaporation. (Dark arrows) Evaporation currents (current
magnitude proportional to arrow thickness). (Gray arrows) Horizontal and rising moist air
currents. (Pale arrows) Horizontal and descending dry air currents.

completely for river runoff. The biotic pump of such ecosystems is not fully functional, and precipitation diminishes exponentially with distance from the coastline.

10.3   THE THEORY OF BIOTIC REGULATION AND THE STABILIZING ROLE OF BIOTA

The energy and substance cycle at the level of biogeocenosis—The biotic community as a mechanism for the maintenance of environmental homeostasis—Biotic regulation and Le Chatelier's Principle—The flux of organic and inorganic carbon in the biosphere— The function of biota's excess productivity—The biosphere as a "biotechnology market"—The biota's threshold of sensitivity to environmental disruptions—The biotic regulation mechanism and ecological succession—The role of repair species in the development of succession—Stabilizing selection as a measure against the decay of biota's genetic memory—Information flux in biota and in civilization—Human capacity for artificial environmental regulation

In all likelihood, Section 10.2 already makes clear the level of organization introduced by biota into its non-living (abiotic) environment. Another important point to be distilled from the foregoing is the universal role of biota in the formation of all the components making up the biosphere. It is difficult not to be amazed by the equilibrium and precise fine-tuning of this grandiose global mechanism capable of unifying highly complex biogeochemical and hydrological cycles, which facilitates the maintenance of the environmental parameters required for life. Now, given that this harmonious coordination characterizes the whole, we can easily deduce that it all begins on the very smallest level: the ecosystem and an individual biogeocenosis: that is, the intracorrelated biospheric unit where every species within a biological community participates in complex trophic chains that serve as conduits for the circulation of energy and chemical substances essential for life.

   If we take energy as a basis, then, as is well known, this cycle begins with the plants (producers). They are the only living organisms—apart from a few bacteria species—capable of synthesizing complex organic molecules out of simple mineral non-organic compounds—thanks to the visible range of the solar radiation spectrum (photosynthesis).[13] It is through plants that the biotic community receives the flow of energy and organic substances which are subsequently used by consumer organisms of the first, second, and subsequent orders (phytivorous animals, carnivores, and detritophagous animals) and finally followed by reducer organisms—bacteria and fungi that are responsible for decomposing dead organic matter. From this view-point, every species occupies a particular ecological niche encompassing not only its

---

[13] Apart from plants (photoautotrophs) there are also bacteria (chemoautotrophs) capable of drawing energy for organic synthesis from the disintegration of certain chemical compounds— hydrogen sulfide, ammonia and others. However, the role of these bacteria in the general balance of the substance and energy cycle in the biosphere is relatively insignificant.

physical location but also its role in the community (its so-called "profession"), its mode of nutrition, and its relationship to other species.

Thus, the roots, trunk, and crown of a given tree shelter several hundred phytivorous insect species, along with their larvae, which feed on the leaves, bark, and underlying layers of wood. In their turn, these animals are the prey of carnivorous insects, birds, and other insectivores. Furthermore, flying insects pollinate the tree during the blooming season while birds pick off the ripe fruit, thereby spreading the seeds through their excrement. When the above-mentioned actors complete their life cycle, the forest's "cleanup crew" steps in. These are insects and other small invertebrates that feed on carrion or use it as a location for laying their eggs. In their turn the larvae coming out of these eggs feed on the same offal. Finally, in the last stage of these transformations, the remnants of undecomposed organic matter are processed by the so-called "true reducers" (i.e., the fungi and bacteria that break down this material all the way to low-molecular compounds). The latter are accessible for assimilation by the root systems of plants, which is how chemical elements essential for life are recycled into circulation.

Naturally, the success and effectiveness of this cycle would be impossible if the scale of consumption by each species were not balanced against consumption by all the other species. Thus, for example, if the activity level of those located higher in the trophic chain were an order of magnitude below that of the bark eaters, aphids, and other so-called "pests" which often reproduce exponentially (the offspring of only one aphid can reach 1024 individuals in the 13th generation by the end of the summer), the entire forest all the way to the treetops could be completely gnawed down in a matter of weeks. On the other hand, the same forest would turn into a foul-smelling ossuary if the detritophagous and reducer organisms could not keep up with the task of processing constantly accruing organic matter.

However, apart from considering the balance within a given community, the ecologist is no less (and perhaps even more) interested in the interaction of the community with its non-living environment. At this level we can temporarily forget about the existence of separate species (the way a zoologist observing the behavior of an animal can dispense with the functioning of its heart and kidneys) and approach a biological community as an autonomous functional unit. Of interest here are first of all not particular but common traits independent of geographic location and specific conditions of habitat. This approach is known in ecology as "ecosystemic", and one of its central objectives is to determine the fundamental aspects equally characteristic of all ecosystems—even those as different from one another as a tropical rainforest, the steppes of southern Russia, or the circumpolar Canadian tundra.

At the most basic level, let us consider a concept common to biology and physics: work. This can be applied to a living organism instead of a machine and is required for a qualitative assessment of the way energy is used and transformed in the process of carrying out various living functions. In this sense the term work can be applied to intracellular synthesis, the transfer of substances from one part of an organism to another, as well as the transmission of an impulse along a nerve fiber, not to mention the mechanical contraction of muscles and the movement of a body in space.

As has already been mentioned, this process begins with the transformation of energy (i.e., from plants capable of capturing energy directly from sunlight). Other living things, on the other hand, can obtain energy from food in the form of intrachemical connections between complex organic molecules. In this respect one could compare not only a given organism, but an entire biotic community to a mechanism that uses energy and nutrients for carrying out a certain type of communally conducted work aimed at maintaining the physicochemical environmental parameters required for life. For example, here is Eugene Odum's description of the manner in which the homeostatic concentration of calcium (a less common biogen) is maintained within a given area of the forest under observation:

> "Retention by and recycling within the undisturbed forest proved to be so effective that the estimated loss from the ecosystem was only 8 kg per ha per yr of calcium (and equally small amounts of other nutrients). Since 3 kg of this was replaced in rain, only an input of 5 kg per ha would be needed to achieve a balance, and this is thought to be easily supplied by the normal rate of weathering from the underlying rock that constitutes the 'reservoir' pool" (Odum, 1971, p. 95).

Referring to other researchers, Odum adds that sod acts as a calcium pump, counteracting the movement of calcium into the depths of the soil and thereby maintaining this biogen in a state of constant cycling.

At the same time, the examples in Section 10.2 demonstrate that the capacity to maintain life-friendly conditions characterizes not only individual communities, but also the Earth's biota as a whole—on the scale of entire continents, the World Ocean, and the whole biosphere in general. This can certainly be viewed as a key aspect of biota's work—at least in terms of global consequences. This brings us to the essential property of all living things: their ability to carry out specific work that maintains the homeostasis of the environment.

The promulgation of conditions conducive to life implies maintenance of an effective resistance against those forces that are always ready to overwhelm life or at least usurp part of its space. Each time these forces are unleashed, a *disruption* occurs (to use a term from the theory of unbalanced systems). Such biospheric disruptions include sharp drops in temperature (glaciations), shifts in the concentration of chemical substances essential for life, hurricanes, floods, etc. To be sure, biota cannot influence such natural phenomena as volcanism, the tides, or changes in the relief of the Earth brought about by tectonic activity, etc. However, biota can compensate for the unfavorable consequences of such events and processes by shifting the balance of consumption by a community toward neutralization of the disruption. This ensures the return of the environment to the undisrupted state, which would be analogous to H. L. Le Chatelier's principle of thermodynamically stable physicochemical systems.[14] Since the main tool in biota's influence over the environment is the synthesis of organic substances from inorganic materials and the decomposition of organic substances into their inorganic elements, we are

---

[14] The system reacts so as to relieve the externally imposed stress.

concerned here with the changes that take place in the balance between the reserves of these substances in the biosphere (Gorshkov et al., 2000).

Thus, an excess of CO_2 in an environment external to a given biotic community can be fixed by means of increased organic synthesis and transformed into low-level forms of organic carbon. At the same time, a deficiency of CO_2 in the air can be compensated for by decomposition of previously formed organic reserves that are located in the soil humus, peat, and organic substance dissolved in the ocean (oceanic humus) where more than 95% of all biospheric organic substance is concentrated. In this connection, the capacity of biota to create high local concentrations of biogens in its environment, which are 100% or greater above biogen concentrations in an external (lifeless) environment, proves clearly that the flow of organic substance synthesis and decomposition exceeds many times the physical flow of biogen transfer. Therefore, in comparison with lower ground layers devoid of living organisms, the soil is highly enriched with organic matter, as well as inorganic compounds essential for the growth of plants. And so it appears that local biogen concentrations in the soil are regulated biotically.

This also pertains to the World Ocean's phytoplankton that fix excess CO_2 originating in the ocean depths (oceanic biotic pump). Therefore, here too, we find the same constructive role played by biota that maintains the CO_2 concentration gradient an order of magnitude higher than if it were determined exclusively by abiotic physical factors. The latter would mainly be temperature differences between the surface and deep oceanic layers that tend to mix very little (Gorshkov et al., 2000). In this manner, by fixing the CO_2 dissolved in seawater, biota prevents this gas from freely diffusing into the atmosphere. This ensures that the concentration of CO_2 in the atmosphere is stabilized at a level acceptable to life.

Another, even larger scale source of inorganic carbon in the atmosphere is volcanism. It has been calculated that the power of this geophysical flow is about 0.01 Gt C/yr. At the same time, the global reserve of atmospheric carbon is of the order of 10^3 Gt C (Degens et al., 1984; Holmen, 1992). Therefore, transfer from the bowels of the Earth could have caused this amount to accumulate over a time period in the order of 100,000 years. However, life on Earth has existed for hundreds of millions of years, and so just in the Phanerozoic (the past 800 million years) the total amount of inorganic carbon in the biosphere would have grown 10,000 times had there not been counterbalancing removal processes such as deposition of carbon in sedimentary rock.

The starting point of this process is rock weathering in which, as was recently established, a key role is played by plants and microorganisms (Schwartzman and Volk, 1989). As was mentioned above, in the weathering process, CO_2 dissolves in rain and groundwater (which is accompanied by the formation of carbonic acid) and reacts with silicate minerals in rocks. It then flows out to the World Ocean in the form of bicarbonate ions. At this stage, after a series of transformations involving sea biota (which then dies off), the carbon is taken out of circulation in the form of organic compounds, settling down as bottom sediments. In some places the thickness of these sediments can reach tens of kilometers, and, according to research conducted in the past few decades, the sediments contain of the order of 10^7 Gt C in the form of

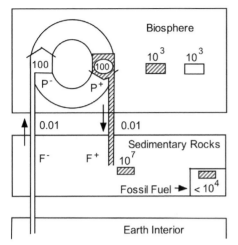

Figure 10.7. Stores (gigatons of carbon) and annual fluxes (gigatons of carbon per year) of carbon in the biosphere shown to an accuracy of the order of magnitude. Stores of carbon are given by figures above small rectangles. Fluxes of carbon are given by figures near arrows. Fluxes and stores of organic carbon are shaded. Fluxes and stores of inorganic carbon are represented by empty areas. The flux of organic carbon going to deposits in sedimentary rocks is equal to the difference between its synthesis and destruction in the biosphere, to an accuracy of about 10^{-4} when averaged over the Phanerozoic (about 6×10^{8} years). Fluxes of synthesis and destruction coincide with one another to the same level of accuracy when averaged over the last several hundred years. That system provides stores of carbon in its organic and inorganic forms in stable states in the biosphere.

dispersed granules accumulated over approximately 1 billion years (Budyko *et al.*, 1987). In other words, the flow of organic carbon deposition in sedimentary rock coincides with the geophysical flow of inorganic carbon into the biosphere (0.01 Gt C/yr) with a relative level of precision in the order of 10^{-4} (Figure 10.7; Gorshkov *et al.*, 2000).

On the other hand, we note the coincidence between the respective orders of magnitude characterizing the global reserves of organic and inorganic carbon in the biosphere. This indicates that biota is capable of maintaining the balance of flow in organic synthesis and organic destruction with great precision. To be sure, we still don't have the means to measure these quantities directly with a sufficient degree of precision, so we can only roughly estimate them to an order of magnitude (102 Gt C; Wittaker and Likens, 1975; Holmen, 1992). Nevertheless, the trends characterizing changes in these reserves in the past and the present can be assessed on the basis of indirect measurements. Thus, for example, research on CO_2 content in air bubbles taken from ice core samples in the Antarctic and Greenland from varying depths (and therefore corresponding to different ages) indicates that carbon concentration in the atmosphere has remained within a narrow range over the past 10,000 years (Rapp, 2008). And over the past million years the concentration of carbon in the atmosphere has remained within the range 180 ppm to 300 ppm (Rapp, 2008). This is of course no coincidence, testifying to the enormous potential of natural biota in compensating for disruptions in the environment in order to maintain life-favoring parameters.

At the same time, one can use the demonstrated ratios (Figure 10.7; Gorshkov *et al.*, 2000) to estimate the rate at which biota processes all the organic and inorganic carbon of the biosphere through synthesis and decomposition. The ratio of the reserves of organic and inorganic carbon ($\sim 10^{3}$ Gt C) to the productivity of the global biota ($\sim$100 Gt C yr^{-1}) characterizes the time of the biological turnover of the biosphere's biogenic reserves, which is of the order of tens of years. That is to say, given only the synthesis of organic substances, all of the biosphere's inorganic carbon would be used up and turned into organic compounds in just a few decades. The converse is equally true: that is, given nothing but decomposition, all of the biosphere's organic carbon would be used up and vanish in merely a few decades. In this connection one may well ask why biota needs this enormous and seemingly excessive biological productivity. After all, in order to compensate for disruptions (e.g., inorganic carbon emanating from the bowels of the Earth) biota could adapt with four orders of magnitude less productivity. However, geophysical processes on Earth are not constant. Along with more or less ordinary environmental disruptions, our geological history has experienced much more serious cataclysms, such as great glaciations, sharp increases in volcanic activity, or impacts from large asteroids. Therefore, the huge capacity for production developed by biota can be interpreted as an adaptive strategy aimed at "stocking up"—just in case of an extreme deviation from the norm. This allows biota to rapidly compensate for extreme environmental disruptions, thereby ensuring the survival of most biological species, as was the case, for example, during the last glacial period.

On the other hand, biota's tremendous capacity also constitutes a certain implicit threat for the environment since the latter can undergo sharp changes within a few decades from a disruption of the balance between synthesis and decomposition. This is possible, for example, as a result of profound changes affecting biota's inner structure—in particular when the species of natural biological communities cease to be intercorrelated. The artificial transformation of nature in a given territory or any attempts to maximize the productivity level of artificial agrocenoses can lead to far greater disruptions of the environment, as well as its accelerated degradation, than the complete annihilation of biota, as in the case of desertification. This is indirectly attested to, among other things, by the rapid increase of CO_2 concentration in the atmosphere over the past 100 years, which until recently used to be perceived only in association with the burning of fossil fuel. It would seem that in response to such an environmental disruption, biota should react in accordance with Le Chatelier's Principle by absorbing excess CO_2 in the atmosphere. However, the global analysis of land use has demonstrated that on significant reclaimed territories, the amount of organic carbon accumulated by land ecosystems does not increase, but rather decreases (Houghton et al., 1983, 1987; Houghton, 1989). Furthermore, the order of magnitude of the rate of carbon emissions into the atmosphere from disrupted continental biota and organic soil reservoirs is about as large as the rate of fossil carbon emissions from the burning of coal, oil, and gas (Watts, 1983). The consequences of the foregoing for the modern biosphere, as well as the resulting theoretical and practical conclusions, will be dealt with in more detail in Chapter 15.

The question is how biota copes with its grandiose mission of stabilizing the planetary environment and how it maintains the synthesis and decomposition of organic substances in such a state of equilibrium for thousands of years and even entire geologic periods. After all, the incredible complexity of life—expressed first of all in the high degree of an organism's intracorrelatedness at the biomolecular, cellular, and organismal levels—is also its Achilles' heel. The more complex the organization level of a given system, the more the system is susceptible to the gradual accumulation of disorder (entropy), which makes the prospect of its degradation and disintegration all the more inevitable. Such disintegration amounts to death for the individual. However, even an organism's genetic code—which in a way guarantees the reproduction of life in a consecutive sequence of generations—is also vulnerable to the accumulation of destructive changes. The latter manifest themselves in the progeny of each separate individual, and the proportion of such faulty individuals is a qualitative trait of a species. For example, in humans 1 out of every 700 newborns is likely to fall victim to Down's Syndrome while 1 out of 100 people living to age 55 manifests the symptoms of schizophrenia (which may have a genetic basis) (Adrian *et al.*, 1985; Sandberg, 1985).

A still more complex form of intracorrelatedness is observed in a biogeocenosis or community of organisms. Here each of the species consuming the main part of the energy flow occupies an ecological niche used by no other species and carries out its own specific part of the work for the stabilization of the environment. Such intracorrelatedness may be very rigid. For example, each of the over 10,000 lichen species constitutes symbiosis between a certain kind of alga and a specific fungus (Farrar, 1976). Some insects can feed only on a single plant species (Raven and Johnson, 1988, 1998).

Therefore, in a normal, undisrupted community, interspecies competition is absent; furthermore, given an almost completely enclosed substance cycle, virtually no waste is produced. In fact, the existence of biologic communities is associated with the necessity of maintaining a highly enclosed substance cycle and the possibility of deviating from that enclosedness in case of external disruptions. On the other hand, in those instances when the supply of nutrients and waste disposal are of an artificial nature, communities come apart. Thus, thanks to their man-made habitat, city-dwelling sparrows can flourish quite successfully outside of any communities, maintaining the stability of their species for thousands of years and experiencing no need to form communities with other city-dwelling birds.

However, if the rigid intracorrelatedness of a biogeocenotic community (local ecosystem) stems from the need to maintain equilibrium between organic synthesis and decomposition, that same characteristic is likely to be the cause of the community's relative transience and its inevitable disintegration. The disintegration of a community, as was stated earlier, is accompanied by the accumulation of mutant individuals that increasingly deviate from the norms of a given species. The other symptom is the gradual weakening of the intercorrelatedness characterizing the relationship among species. The latter struggle for food resources and begin to occupy intersecting ecological niches. Such a community is no longer capable of maintaining the stability of the local environment, losing its competitive ability and eventually disappearing from the face of the Earth.

This contradiction between the relative immortality of life as a whole, on the one hand, and, on the other hand, the finite nature of a separate individual, as well as that of a "superorganism"—biogeocenosis—is resolved by nature in a rather uneconomical, but the only possible manner. We are referring to competition and natural selection of independent individuals within the same species, as well as biological communities within an ecosystem. In this connection survival within a given population is possible only for those organisms in possession of an undisturbed genetic code or those communities characterized by an optimal species structure. Such characteristics make it possible to successfully compensate for random environmental fluctuations and disruptions. Similar competitive interaction forces ineffective producers out of the free economic marketplace, and by the way, it is no coincidence that a section in *Physical and Biological Bases of Life Stability: Man, Biota, Environment* by Gorshkov is entitled "The biosphere as a free market" (Gorshkov, 1995).

Let us recall the firestorm of criticism unleashed against J. Lovelock by biologists and evolutionary scientists. In this connection, we should consider the key differences between the Gaia Hypothesis and V. Gorshkov's theory of biotic regulation. Lovelock sees a grandiose global mechanism, unifying into one indivisible whole, the living and non-living components of the biosphere for the purpose of preserving the vital parameters of the planetary environment. Gorshkov introduces a "biotechnology marketplace" formed by a great multitude of biological "subjects"—communities and individuals. Lovelock describes the colossal complexity of the substance and energy currents permeating the biosphere and maintained over tens and hundreds of millions of years. He stresses that these currents start out as unstable and are condemned to inevitable spontaneous disintegration. Gorshkov emphasizes the necessity of constantly affirming one's "ecological competence" and the right to a place in the sun for every individual, as well as every local community, passing through the sieve of competitive selection. The latter process reinforces the winners' genetic and species structure in its progeny. Therefore, according to Gorshkov, nature establishes order by working with countless individual operational units—on the basis of the statistical law of large numbers—thereby minimizing random fluctuations that threaten the existence of any complex intracorrelated system.[15]

It ought to be said that such adaptive phenomena can be encountered at the level of individual organisms as well. For example, the "distributive" circulatory system in animals uses numerous uncorrelated blood corpuscles: erythrocytes for ensuring a reliable supply of oxygen to cells and leucocytes for neutralizing elements foreign to the organism. Another example is the chaotic mass of randomly oriented leaves on most trees and bushes. This material is responsible for the lion's share of the organic production in the majority of ecosystems.

However, perhaps the most interesting (and important) element in Gorshkov's conception is the notion of biota's sensitivity threshold when it comes to environmental disruptions. As in the case of the Gaia Hypothesis, the theory of biotic regulation was also harshly criticized by neo-Darwinists. However, the grounds for this criticism were quite different. As has been noted earlier, classical evolutionary

---

[15] By the way, many large companies have recently begun to follow this principle. They prefer structures with relatively autonomous units characterized by weak ties to the central administration. Such units can act unhindered by the deadweight of centralization.

theory has at its core the fate of the individual, which is why this theory has been jokingly accused of individualism. However, most jokes have an element of truth, and this is no exception. Indeed, according to classical evolutionary theory, natural selection deals with different variations in individuals within the same population (species) adapting with varying degrees of success to changing environmental conditions. Successful adaptations increase the chances of survival and the transmission of one's genotype through progeny. Those whose adaptations are adequate for the changed habitat don't really "care" about the precise nature of a given adaptation (i.e., whether or not this adaptation may endanger the well-being of the whole population or ecosystem). The individual is just "happy" to survive at a given moment in time and live within a set of concrete changing circumstances.

Things are totally different when the selection criterion is the capacity to carry out the work required for the stabilization of the environment. It would seem that natural selection should not differentiate between communities and individuals within them on the basis of how well they accomplish this mission. After all, whether environmental conditions deteriorate or improve, communities and individuals are in the same boat. Furthermore, communities working for the "common good" and expending part of their energy for the purpose should be outcompeted by those communities that can save this energy. So how do they keep their place in the sun? And why has life's ability to carry out biotic regulation not deteriorated over billions of years, turning into an endless series of new species formations? Instead, this capacity has been passed on from generation to generation. In order to answer this question, V. Gorshkov proposes the notion of biota's sensitivity to environmental disruptions—a faculty that allows it to react only to those changes that do not exceed a certain threshold (Gorshkov et al., 2000).

The reader may have noticed that in the summer under the thick foliage of certain trees in a park or a forest one tends to breathe differently than in an open, wind-swept meadow. The same applies to humidity and temperature differences. In other words, within the limits of their crown, trees can maintain a kind of microclimate or homogeneity of soil moisture content over the area of their root distribution system (Gorshkov and Makarieva, 2006). Soil specialists have noted something else as well: as we move from tree to tree (even of the same species), we see that the soil profile is often characterized by a border visible to the naked eye—in terms of color, structure, and texture. And yet, it has been long established that soil formation results from the coordinated work of all the elements in a biogeocenotic community, including microorganisms and fungal flocci. So if the results of this work can vary even among sites covered by neighboring trees, every one of these communities acts as an independent unit in the biological cycle.

Of course one cannot compare the local environment on the scale of one tree and the correlated biota (in and above the soil: underbrush, bacteria, fungi and small invertebrates) with an animal's internal environment that is maintained in a homeostatic state. The tree's environment is swept by winds and washed by rain, which is why various fluctuations are practically inevitable in this case. Furthermore, atmospheric mixing and the physical diffusion of biogens in the soil appear at first glance to eliminate any possible distinctions in a local microcosm of biotic communities. And yet, the capacity of biological productivity on the part of certain mature trees is such that their crowns can form an internal atmosphere where, for example, the CO_2 content can differ by a few percent from the average atmospheric concentration.

The issue is how biota itself regulates such differences. Calculations enable us to conclude that biota's sensitivity to changes in most environmental parameters is in the range 10^{-2}–10^{-3} (Gorshkov *et al.*, 2000). In this case, if a shift in atmospheric CO_2 concentration unfavorable to a given community equals less than 10^{-2} (e.g., 1/1000 of a percent), biota will not notice this change and will not react to it. On the other hand, a 1% difference in CO_2 concentration can turn out to be critical for a normal community and can lead to its functional restructuring. This restructuring could, for example, take the form of depositing excess carbon in the organic humus of the soil, or—given an insufficiency of atmospheric CO_2—the result could be an increase in the processes of destruction and liberation of inorganic carbon. The same applies to soil moisture content that can be influenced within certain limits by individual trees; in this manner moisture content can be raised or lowered (through changes in transpiration, the vertical temperature gradient below the forest canopy, emissions of biogenic aerosols into the atmosphere, etc.; Gorshkov and Makarieva, 2006). And so, communities with a sensitivity level of $<10^{-2}$ acquire a small but palpable degree of superiority with respect to mutants incapable of maintaining the local environment in a state favorable to their needs. As a result the former are more likely to survive than the latter in the process of individual natural selection with the losers gradually ending up displaced from the ecosystem.

And now let us imagine the following scenario: as a result of certain destabilizing influences (e.g., volcanic eruptions of CO_2) the concentration of this gas in the atmosphere significantly exceeds values optimal for biota. In that case, given a corresponding sensitivity level of biotic communities making up a certain ecosystem, the inorganic carbon entering the ecosystem will become fixed, shifting to an inactive organic state. If the total area occupied by such an ecosystem is sufficiently large, we will witness a globally significant physical biogen flux from the external environment into the realm of life's functioning. It is clear that this flux will continue to exist until the respective CO_2 concentrations within the bounds of the ecosystem and beyond it approach equality to one another with a degree of precision matching biota's corresponding sensitivity, in other words—when the global concentration of the biogen in the external environment reaches a level favorable to biota. This is the basic scheme of the biotic regulation mechanism passed on from generation to generation in the process of competition and selection affecting individual biotic communities.

An argument in favor of environmental biotic regulation can be made in terms of the phenomenon of *ecological succession*—discovered by H. Cowles (1869–1939) and studied in great detail by F. Clements (1874–1945). This is a process of ecosystem development characterized by distinct stages whereby dominant species replace one another in a seemingly almost programmed manner. For example, a newly formed volcanic island in the middle of the ocean is settled by cyanobacteria and pioneer communities of lichens that do not require soil cover. Ten or more years will likely pass before these organisms form the soil cover needed by more complex organisms. At first the latter could be mosses and ferns, then meadow plants, followed by scrub and finally—trees. Every preceding community brings the next one along "by hand" as it were, passing on the relay baton and yielding its place. At the final stage of succession we see the appearance of a community that—in the absence of external disruptions—is self-sufficient and stable enough to maintain a state of equilibrium with the environment for an indefinitely long time given the constancy of the biomass and density characterizing the specific constituent species. Such apex formations— known as climax communities—include oak forests on moist clay soil or boreal pine forests on sandy soil and spruce forests on loamy soil (the latter two being typical of Northern Europe).

The above settlement sequence on a naked volcanic surface illustrates the notion of primary succession. However, similar stages are observed in the process of secondary succession (e.g., during the regeneration of a coniferous forest after a severe impact such as a hurricane, logging, or a fire). Thus, 30 years after a forest fire, the affected area will be characterized by a highly chaotic vegetational cover and maximal entropy in the distribution of productivity among individual bush and tree species. During this time trees grow with maximal speed; their wood is sterile and does not take part in metabolic processes. The forest regenerates itself and is practically incapable of regulating the environment. The unenclosedness of the substance cycle in such communities can reach tens of percent (Gorshkov, 1980). However, after a few decades following the disruption, this unenclosedness—as demonstrated by measurements of productivity, biomass accretion and changes in the concentrations of inorganic soil substances—dwindles down to a few percent (Bormann and Likens, 1979). By 50–70 years after the disruption, the productivity of the disrupted community becomes stabilized. The same goes for the leaf surface and, for the most part, the enclosedness of the biogen cycle with maximum biogen concentration in the upper soil horizons. Finally, 150 years after the disruption, most of the community's characteristics return, including its biomass, the thickness of the forest floor cover, the content and distribution of chemicals in the soil, as well as the integrity of the biogen cycle. The complete recovery of the forest after logging or a fire—the formation of the tree layer's multi-age structure—takes place only after 200–300 years have passed (Finegan, 1984).

The chemical composition of the environment also undergoes significant changes in the course of succession. This affects the soil in the first place. Local concentration levels of individual biogens can change tenfold to a hundredfold, depending to a large extent on the life activity of the species that determine the direction of succession change. Such organisms are referred to by V. Gorshkov as "repair species" in accordance with their role in an ecosystem undergoing regeneration after a disruption. Typical repair species in boreal coniferous forests include, for example, birch, alder, aspen, berry plants, pileate fungi, and most animals feeding on these species. The most notable characteristic of the repair species is their ability to shift the concentration of environmental nutrients in a direction unfavorable to themselves but favorable to the next generation. This explains the phenomenon of succession stages, namely, the displacement of a repair community dominant at a given point in time and its replacement with the next repair generation. The latter finds temporary conditions optimal for its needs but subsequently vacates its spot to the next dominant group.

In the final stage of succession, the biogen concentration of a local environment reaches values favorable for climax species and unfavorable for all repair species. In this manner, the disrupted community returns to its initial, stable state that is characterized by certain typical traits. The latter include the accumulation (by the time succession comes to an end) of more and more available nutrients in the community's biomass and the simultaneous depletion of biogens in the abiotic components of the system (i.e., water and soil minerals). The amount of detritus increases. This waste becomes a key source of nutrients in ecosystems as detrivorous organisms take the role of the system's main consumers away from the herbivores, etc. (Green *et al.*, 1984, Vol. 2). Under such conditions it is the climax species that gain the best competitive advantage, forming a stable population capable of maintaining this advantageous state for a long time. As for the repair species in a climax community, they find themselves hampered by an unfavorable environment and continue to exist only as separate marginal individuals that do not form a population. This process goes on until the next cycle.

Such is the general outline of the succession process that appears to be very specific for each type of climax community and practically independent of the latter's geographic location. However, all this applies only in the absence of new disruptions that can hamper succession, or cut it off completely. Then the whole process returns to its starting phase. If disruptions recur repeatedly, the consequence may be irreversible damage to the ecosystem that ends up "forgetting" the program for its regeneration and cannot return to the climax phase. This happens, for example, as a result of regular logging after the formation of so-called "business wood" (suitable for economic exploitation) or when forests are treated with chemical pesticides (for the selective suppression of less valuable tree species). Another scenario is the artificial thinning out and clearing of mature forests in order to eliminate declining trees, as well as fallen or rotten trunks. The latter, by the way, is an extremely ruthless and dangerous disruption of a natural community's life because it is precisely the mature forest that constitutes the healthiest biospheric body. Here we find a perfect balance in the substance cycle, and therefore there can be nothing superfluous: leaf accretion is limited by leaf consumption on the part of small invertebrates; wood and root accretion is limited by the action of fungi and bacteria. In general, all the organic components of both sturdy and declining trees are used for the life needs of other organisms. In this manner, the widespread practice of periodic logging at 50-year intervals ends up cutting short the regeneration process of primary climax forests with an enclosed substance cycle and a capacity to compensate for environmental disruptions (Houghton, 1989). Therefore, to achieve a return to an undisrupted state of the biosphere, the intervals between successive loggings must be increased to at least 300 years, thereby reducing the rate of logging by a factor of 6. And if we take into account the fact that clear-cutting today exceeds the volume of natural accretion—which leads to the progressive diminution of forested areas—forest harvesting has to decrease globally by at least 10 times (more in Chapter 15).[16]

It is clear that the above stage-like structure of succession could not have been replicating itself for thousands of years if it were not fixed in the genetic memory of biota (i.e., in the genome of every species). For example, all the repair species in the framework of a given succession are genetically programmed to change the environmental parameters to a direction unfavorable to themselves and favorable to the climax species. However, keeping in mind the special role of the climax species in the maintenance of biospheric stability, we can easily understand that these changes in environmental parameters are beneficial not just for biota as a whole but in the end for the repair species as well. Therefore, the capacity of a climax community to maintain environmental stability is inseparable from the corresponding genetic information and an appropriate set of natural species. However, over the course of time, genetic memory, like any other ordered information, is subject to gradual disintegration and decay. And so, if we consider the capacity of biota to maintain certain environmental conditions, we have to note the mechanism that ensures the intactness of this genetic program as it is passed on from generation to generation. This mechanism, as has been pointed out before, is competition and the natural selection of individuals, as well as their communities.

Evolutionary theory distinguishes between several types of natural selection, depending on the goals placed before a given population by changing environmental conditions: stabilizing, directional, disruptive, etc. We are talking specifically about stabilizing selection that is aimed at conserving average phenotypical traits and thereby ensuring the adaptation of a given population to a familiar habitat. By eliminating individuals with extreme phenotypical deviations, stabilizing selection (at the population level) blocks the disappearance of genetic information as a result of random mutations. This maintains systemic order and prevents the accumulation of entropy in the system. However, in a way the natural selection of individuals is a means of measuring their quality (i.e., the degree of their suitability for carrying out a given biological job). And like any process of measurement, this one must possess a minimum resolution capability (e.g., reacting to mutation change in the genome only

---

[16] To put the damage caused by deforestation and the associated loss of ecosystems into economic perspective, we would like to cite an ongoing study entitled *The Economics of Ecosystems and Biodiversity* (TEEB). The research is being conducted under the patronage of the German Federal Ministry for the Environment and the European Commission. According to TEEB's leader Pavan Sukhdev, the harm from the current financial crisis is far less daunting than the yearly losses from deforestation. In an interview with BBC News, Sukhdev said: "So whereas Wall Street by various calculations has to date lost, within the financial sector, $1–$1.5 trillion, the reality is that at today's rate we are losing natural capital at least between $2–$5 trillion every year" (October 10, 2008). Expressing our self-destructiveness in the language of money brings two seemingly incompatible fields—economics and ecology—together. It turns out that the scientists and the money movers are dealing with two aspects of the same phenomenon. And this underscores the notion of the biosphere as a truly all-encompassing entity: nothing is outside of it—neither the trading floor nor the forest floor.

above a certain threshold). In this light, individuals with a clearly altered genetic program and manifest gross deviations that lead to a decrease in their competitive ability are displaced from a given population. Others, manifesting changes below this threshold, are successful at surviving through the biological thresher and creating the basis for intraspecific genetic diversity.

The above makes it possible to attempt an explanation of the mystery constituted by the phenomenon of *species discreteness*—a stumbling block for the classical theory of evolution. If the course of evolution is uninterrupted in time, with species constantly adapting to changing environmental conditions, how can we explain the absence of transitional forms? This applies to modern material, as well as paleontological data. It all falls into place given the above interpretation of stabilizing selection that ignores insignificant phenotypical deviations, but takes effect when faced with changes above a certain species threshold. At the same time the existence of this threshold offers us a key to understanding the amazing constancy of species— a phenomenon whose duration can be compared to geological periods.

And yet there are situations where stabilizing selection moves to the background, as it were, yielding its place to other natural selection mechanisms. This happens, for example, when biota exhausts its regulatory capabilities at critical stages of its historical development. As noted earlier, there is a whole series of abiotic processes in the bowels of the Earth, as well as in space, which cannot be influenced by biota. A striking example is the transformation of a reducing atmosphere into an oxidizing atmosphere that took place about 2.2 billion years ago. As a result, the biosphere was "turned inside out" and became mainly nitrogenous–oxygenic with a few oxygen-free pockets where anaerobic microorganisms found refuge (Zavarzin, 2001, Kirschvink and Kopp, 2008). This was brought about by the process leading to the formation of the planet's central core where most of the iron ended up through gravitational forces. Consequently, the concentration of ferrous iron (FeO) in seawater dropped drastically. While in the previous 1.5 to 2 billion years all of the oxygen formed by life activity of anaerobic prokaryotes was used for the oxidation of atmospheric gases (NH_3, CH_4, CO, and H_2S) and the ferrous iron dissolved in the ocean, now the freed oxygen started to accumulate in the atmosphere. This affected almost all the prokaryotic biota that, as a whole, was not adapted for life in an oxidizing environment. As a result of these catastrophic events on Earth, a global change in the composition of biota species took place, and the role occupied up to that point by the predominant anaerobic microorganisms was replaced by the relatively few (at the time) photosynthesizing bacteria. The latter used water and CO_2 to build organic molecules, and their energy source was the visible solar spectrum.

However, even incomparably smaller scale environmental transformations, accompanied by mass extinctions of old species, have not occurred very often in the history of the planet. Thus, in the past half a billion years, this has happened on average only once every 100 million years (Raven and Johnson, 1988; Jablonsky, 1994). Furthermore, it took millions of years for a species turnover in biota to take place (i.e., entire geological periods). Unfortunately, current global environmental destabilization processes occurring under the influence of human economic activity are comparable to biota's transformations from past geological epochs. The difference in the timescale of these changes, however, amounts to several orders of magnitude. Biota is unable to come up with a response to this pressure: for example, it cannot produce new species out of a hat since such processes require tens of thousands of years according to paleontological data. In this connection, a real threat is posed by the genetic disorganization of existing species and the consequent loss of hereditary memory regarding the biotic regulation mechanism.

Stabilizing selection is truly effective only under conditions existing in the natural ecological niche of every species. Individuals with normal or mildly altered genetic programs are the most competitive. They constitute a population whose genome is the keeper of information regarding those qualities of a given species that ensure environmental conservation. However, when natural habitat conditions are distorted, the competitive capacity of these individuals deteriorates sharply. The advantage now passes on to individuals with a disrupted genome that no longer possesses any memory regarding the optimal structure and numbers of a given species. And yet this memory is essential for maintaining specific environmental conditions. This pertains not only to domesticated animals or cultivated plants that have long lost the connection to their natural origins, but also to many so-called synanthropic species (i.e., those that exist in the vicinity of humans). These species, whose ecological niche has been distorted under the conditions created by civilization, include house mice that are no longer able to return to their natural state, or sparrows whose numbers have increased by several orders of magnitude and who have become very rare beyond zones of human habitation.

Something similar is observed in intensively exploited forests that for all intents and purposes cannot return to their climax phase since the genetic information regarding an environment optimal for climax species has been lost forever. Given the presence of repair species desired by humanity and artificially maintained by it, the community essentially loses its capacity for biotic regulation. And should humanity reclaim the entire biosphere (or perhaps even long before this happens), the mechanism of environmental regulation and stabilization may be lost on a global scale. In this case the "master of the planet" would have no choice but to take over all responsibility for environmental regulation by replacing biotic regulation with its technogenic counterpart. The question is whether this task corresponds to our real capabilities.

Before trying to answer this question, let us compare the information flux passing through biota and through human civilization, respectively. In the final analysis, the magnitude of this flux determines the effectiveness of environmental regulation carried out by both natural and artificial means. Gorshkov *et al.* (2000) compared (1) the volume of biota's genetic information with the volume of human cultural information and (2) the intensity of information flux processed by the organisms of the biosphere and by all the computers of civilization.

Genetic information is encoded in the cell nucleus by means of various combinations consisting of four nucleotide pairs. The human genome contains 3×10^9 nucleotide pairs (bp) (Lewin, 2007), and the magnitude of the coded information is estimated at 6×10^9 bits. On the other hand, calculations indicate that the amount of information stored in human memory corresponds approximately to the volume of our genetic information. Indeed, the human brain can assimilate no more than 100 bits/s of which long-term memory is on average the recipient of approximately 10 bits/s. The first 20 years of our lives constitute the period when new information is assimilated most actively, which amounts to 6×10^8 s. Subsequently, memory volume increases insignificantly: that is, we can assume that the order of magnitude characterizing the information stored in the human brain does not change much after age 20. Therefore, the memory reserves of an adult can be estimated at 6×10^8 s $\times 10$ bits/s $= 6 \times 10^9$ bits. Multiplying 6×10^9 bits by 6×10^9 people, we would estimate about 10^{19} bits as the magnitude corresponding to the cultural information of humanity.

However, as opposed to animal brains, the human brain also carries collective information stored in the brains of millions of people. This collective information is essentially a cultural relay—passed on from generation to generation. In order to estimate the cultural information of humanity, we have to remove the mutually non-overlapping part of that memory stored in the brains of a small class including specialized scientists, authors, artists, musicians, etc. from the collective memory of the 6×10^9 humans living on Earth. As for information recorded in books and computers, when not retrieved by a living member of society, it is essentially dead and lost. Therefore, such information can be discounted as a separate value. Such specialists constitute at most 10% of the population, and each field of knowledge requires at least 100 specialists sharing the same memory information. Therefore we must multiply the previous figure of 10^{19} by 0.1×0.01 to yield 10^{16} bits as the magnitude corresponding to the original cultural information of humanity.

Let us now turn to the genetic information of the natural biota that forms the Earth's biosphere. It can be estimated by multiplying the number of species in the biosphere—10^7 (Thomas, 1990)—by the average number of nucleotide pairs in the genome of one species. The result is of the order of 10^9 bp and can be considered as practically matching the genetic information of the largest species group (namely, the insects). Therefore, the total volume of the genetic information stored in the natural biota is $\sim 10^{16}$ bits, and its order of magnitude corresponds to the volume of the cultural information produced by civilization (see Figure 10.8; Gorshkov *et al.*, 2000). In the same way, the total genetic information of the biota is not superior to the combined memory of modern computers which can be determined by multiplying the average memory capacity of one PC (10^9 bits) by the total number of computers in exploitation. However, life is a process characterized first of all by information flux, as well as the work that can be carried out by living organisms per unit time as they interact with the environment. And if we consider these parameters, an unbridgeable quantitative gap separates civilization and biota.

For example, the power of information flux (metabolism) in an individual bacterial cell (10^8 bits/s) is comparable to the information flux of a modern personal computer. In the case of the bacterium, the logical elements are molecular units while the cell itself acts as the gauge that measures changes in environmental parameters. The transformation of information in a computer takes place almost without any energy expenditure (i.e., with a coefficient of efficiency close to 1), and the same can be

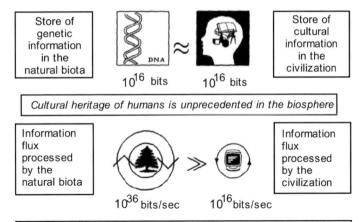

Figure 10.8. Major informational characteristics of natural biota and civilization.

said about the informational processes in living biota. For example, the development of the embryo in an egg—which is a natural "microprocessor"—is accompanied by virtually no influx of external energy. Of significance here is only the temperature that is optimal for different species. A fry or chick egg loses less than 20% of its initial mass while the coefficient of efficiency characterizing the transformation of information in this egg is over 80% (Kendeigh, 1974). In other words, in this case, energy is being used with maximal efficiency. In order to approximate the level of natural "technology", humans would need bacteria-sized computerized devices for interaction with the environment. We would also have to create computational programs equivalent to the genomes of several million species.

We can now juxtapose information flux in the biota (which can be equated with metabolism—Figure 10.8; Gorshkov *et al.*, 2000) with information flux in all the computers of civilization. If we take the average information capacity of an individual computer as 10^8 bits/s, the flux in the computer network would equal 10^{16} bits/s. In biota a bacterium or any other cell can act as an adequate analogue of a computer. The number of bacteria on Earth is estimated at 3×10^{27} while the total number of cells in the biosphere is approximately an order of magnitude greater. We can, therefore, estimate the flux of information processed by the planet's biota at $10^8 \times 10^{28} = 10^{36}$ bits/s. It ought to be noted that this figure is 20 orders of magnitude higher than the flux of information in all the computers run by human civilization. And if we assume that biospheric regulation could someday be taken over by the computers of the future (whose information capacity potential is thought to be up to 10^{12} bits/s), the interaction of these machines with the environment would still be fundamentally different from similar processes on the part of living cells. In the latter case, molecular units of memory are combined with elements that interact with the environment. This applies not only to unicellular organisms (almost all of whose intracellular processes take place in contact with the environment), but also to fungi and the higher plants which retain this characteristic thanks to their large effective surface—extremely thin branching flocci, high leaf area index, ramified root system, etc.

As the above estimates indicate, in the foreseeable future the gap between the information flux of biota and that of human civilization is likely to diminish by six orders of magnitude (i.e., from 10^{20} down to 10^{14}). This will happen thanks to increases in the number of computers and their speed. However, even if we could eliminate this gap altogether, our problem would not solved: that is, we would not be in possession of a technological system for environmental regulation equivalent to what biota has at its disposal. In order to illustrate this point, let us recall the famous conflict between manual and automatic control mechanisms. Manual control is carried out on the basis of inborn and acquired information, as well as peripheral impulses, coming from sensory organs along feedback channels. This type of control is limited by the rate at which the human brain can assimilate this information. Automatic control is based on computer programs and unfolds at a rate millions of time greater than the capacity of the human brain. In the latter case, human beings must be absolutely certain of the program entered into the computer, which necessitates numerous preliminary experiments. And yet, under various abnormal circumstances people are forced to take over from computerized control systems, relying on their own experience, knowledge, and intuition—even if this means slowing down the process in question.

Biotic environmental regulation is a control system where the speed of information processing exceeds human mental capacity by 34 orders of magnitude, and it is 20 orders of magnitude beyond today's computerized control systems. In point of fact, this automatic environmental control system operates on the basis of a program that has been undergoing testing and upgrading for billions of years. Over this period geophysical and extraterrestrial factors have caused Earth biota to undergo significant change approximately once every 100 million years. This involved mass extinctions of old species, as attested by paleontological data. In other words, in the past billion years of the biosphere's existence, no more than ten environmental control programs have been tried out. Each of these programs must have been unique for its time and was maintained by biota for as long as was necessary. The creation and fine-tuning of new biotic programs underwent many thousands of tests in the course of the evolutionary process—without any interruptions in the continuity of life's universal biochemical organization.

The inevitable conclusion is that attempts on the part of humanity to find an adequate replacement for biotic environmental regulation would also require tens if not hundreds of thousands of years since the testing and correcting of such programs would necessarily take place under conditions of manual control. But why should we set such strange goals for ourselves to begin with? Would it not make more sense to give the sophistication of natural biota its due and express our admiration by preserving it, helping it expand, and regenerating as much as possible what has been destroyed in the pointless war between man and nature?

References

Barnola, J. M.; P. Pimienta; D. Raynaud; Y. S. Korotkevich (1991). "CO_2 climate relationship as deduced from the Vostok ice core: A re-examination based on new measurements and on the re-evaluation of the air dating," *Tellus V* **43B**(2), 83–90.

BBC News Online (October 10, 2008). Available at *http://news.bbc.co.uk/2/hi/science/nature/7662565.stm*

Adrian, J. E.; J. M. Allen; S. R. Bloom *et al.* (1985). "Neuropeptide Y distribution in the human brain," *Nature* **306**, 584–586.

Bolin B. (1983). "The carbon cycle," in B. Bolin, R. B. Cook (Eds.), *The Major Biogeochemical Cycles and Their Interactions: SCOPE 21*. New York: John Wiley & Sons.

Bormann, F. H.; G. E. Likens (1979). *Pattern and Process in Forested Ecosystems*. New York: Springer-Verlag.

Broecker, W. S.; D. M. Peteet; D. Ring (1985). "Does the ocean–atmosphere system have more than one stable mode of operation?" *Nature* **315**(6014), 21–26.

Budyko, M. I.; A. B. Ronov; A. L. Yanshin (1985). *History of the Earth's Atmosphere*. Berlin: Springer-Verlag.

Climate Change (2007). IPCC (Intergovernmental Panel on Climate Change), Fourth Assessment Report. Available at *http://ipcc-wg1.ucar.edu/wg1/Report/AR4WG1_Print_Ch03.pdf*

Degens, E. T.; S. Kempe; A. Spitzy (1984). "Carbon dioxide: A biogeochemical portrait," in O. Hutziger (Ed.), *The Handbook of Environmental Chemistry*, Vol. 1, Part C. Berlin: Springer-Verlag, pp. 125–215.

Es'kov, K. Iu. (2008). "The History of the Earth and life on it" [in Russian]. Available at *http://www.paleo.ru/paleonet/publications/eskov/08.html*

Farrar, J. F. (1976). "The lichen as an ecosystem: Observation and experiment," in D. H. Brown, D. L. Hawksworth, R. H. Bayley (Eds.), *Lichenology: Progress and Problems*. New York: Academic Press, pp. 385–406.

Finegan, B. (1984). "Forest succession," *Nature* **312**(8), 103–114.

Gorshkov, V. G. (1980). "The structure of biospheric energy flux," *Botanichesky zhurnal* **65**(11), 1579–1590 [in Russian].

Gorshkov, V. G. (1995). *Physical and Biological Bases of Life Stability: Man, Biota, Environment*. Berlin: Springer-Verlag.

Gorshkov, V. G.; A. M. Makarieva (2006). "Biotic pump of atmospheric moisture as driver of the hydrological cycle on land," *Hydrology and Earth System Sciences Discussions* **3**, 2621–2673.

Gorshkov, V. G.; A. M. Makarieva (2007). "Biotic pump of atmospheric moisture as driver of the hydrological cycle on land," *Hydrology and Earth System Sciences* **11**, 1013–1033.

Gorshkov, V. G.; V. V. Gorshkov; A. M. Makarieva (2000). *Biotic Regulation of the Environment: Key Issues of Global Change*. Chichester, U.K.: Springer/Praxis. Also see V. G. Gorshkov's and A. M. Makarieva's site on questions of environmental biotic regulation at *http://www.bioticregulation.ru/index.php*

Govorushko, S. M. (2006). *Glaciers and Their Importance for Human Activity*. Vladivostok: Pacific Geographical Institute, FEB RAS, Vol. 6, pp. 60–70.

Green, N. P. O.; G. W. Stout; D. J. Taylor (1984). *Biological Science*, Vols. 1 and 2 (edited by R. Soper). Cambridge, U.K.: Cambridge University Press.

Hoffman, P. F.; Daniel P. Schrag (2002). "The snowball Earth hypothesis: Testing the limits of global change," *Terra Nova* **14**(3), 129–155. Available at *http://www.bioticregulation.ru/index.php*

Holmen, K. (1992). "The global carbon cycle," in S. S. Butcher, R. J. Charlson, G. H. Orians, G. V. Wolfe (Eds.), *Global Biogeochemical Cycles*. London: Academic Press, pp. 239–262.

Houghton R. A. (1989). "The long-term flux of carbon to the atmosphere from changes in land use," extended abstracts of papers presented at the *Third International Conference on Analysis and Evaluation of Atmospheric CO_2 Data*. Heidelberg: WMO University, pp. 80–85.

Houghton R. A.; E. Hobbie; J. M. Melillo; B. Moore; B. J. Peterson; G. R. Shaver; G. M. Woodwell (1983). "Changes in the content of terrestrial biota and soils between 1860 and 1980: Net release of CO_2 to the atmosphere," *Ecological Monographs* **53**, 235–262.

Houghton R. A.; R. D. Boone; J. R. Fruci; J. E. Hobbie; J. M. Melillo; C. A. Palm; B. J. Peterson; G. R. Shaver; G. M. Woodwell; B. Moore (1987). "The flux of carbon from terrestrial ecosystems to the atmosphere in 1980 due to changes in land use: Geographic distribution of the global flux," *Tellus* **39B**, 122–139.

Hsu, K. J. (1992). "Is Gaia endothermic?" *Geological Magazine* **129**, 129–141.

Imbrie, J.; K. P. Imbrie (2005). *Ice Ages: Solving the Mystery*. Cambridge, MA: Harvard University Press.

Itelson, L. B. (2000). *Lectures on General Psychology*. Minsk: Kharvest/Moscow: OOO Isdatel'stvo AST [in Russian].

Jablonsky D. (1994). "Extinctions in the fossil record," *Phil. Trans. R. Soc. London* **344**(1), 11–17.

Kasting, J. F. (1993). "Earth's early atmosphere," *Science* **259**, 920–926.

Kendeigh, S. C. (1974). *Ecology with Special References of Animals and Man*. Englewood Cliffs, N.J.: Prentice-Hall.

Khain, V. E. (2007). "Interaction of the atmosphere, biosphere and lithosphere: A key process in the evolution of the Earth," *Vestnik Rossiiskoi akademii nauk* **77**(9), 794–797.

Kirschvink, J. L.; R. E. Kopp (2008). "Paleoproterozic icehouses and the evolution of oxygen mediating enzymes: The case for a late origin of Photosystem-II. *Phil. Trans. R. Soc. Lond., Ser. B.* **363**, 2755–2765.

Koronovsky, N. V. (2002). *General Geology*. Moscow: Izdatel'stvo Moskovskogo universiteta [in Russian]. Available at *http://dynamo.geol.msu.ru/TextBooks/ObGeol/chapter14.pdf*

Krasilov, V. A. (1992). *Natural Conservation: Principles, Problems, Priorities*. Moscow: Institut okhrany prirody i zapovednogo dela.

L'vovitch, M. I. (1979). *World Water Resources and Their Future*. Washington, D.C.: American Geological Union.

Lapo, A. V. (1987). *Traces of Former Biospheres*. Moscow: Znanie [in Russian].

Lewin, B. (2007). *Genes IX*. Sudbury, MA: Jones & Bartlett.

Lovelock, J. E. (1988). *The Ages of Gaia: A Biography of Our Living Earth*. New York: Oxford University Press.

Marfenin, N. N. (2006). *The Sustainable Development of Humanity*. Moscow: Izdatel'stvo MGU [in Russian].

Melekestsev, I.V. (1969). "Volcanism as a possible cause for glaciation," *Volcanos and Eruptions*. Moscow: Nauka, pp. 140–149 [in Russian].

Neftel, A.; H. Oeschger; J. Schwander; B. Stauffer; R. Zumbrunn (1982). "Ice core sample measurements give atmospheric CO_2 content during the past 40,000 yrs," *Nature* **295**, 220–223.

Newkirk, G. (1980). "Solar variability of time scales of 105 years to 109.6 years," in R. O. Repin, J. A. Eddy (Eds.), *Proc. Conf. Ancient Sun, USA*, pp. 293–320.

Odum, E. P. (1971). *Fundamentals of Ecology*, 3rd Edn. Philadelphia: W. B. Saunders.

Rapp, D. (2008). *Assessing Climate Change*. Chichester, U.K.: Springer/Praxis.

Rapp, D. (2009) *Ice Ages*, to be published by Springer/Praxis.

Raven, P. H.; G. B. Johnson (1988). *Understanding Biology*. Wm. C. Brown Publishing, Dubuque, IA.

Raven, P. H.; G. B. Johnson (1998). *Understanding Biology*. St. Louis: Mosby.

Rocha, H. R. da; M. L. Goulden; S. D. Miller; M. C. Menton; L. D. V. O. Pinto; H. C. de Freitas; A. M. E. Silva Figueira (2004). "Seasonality of water and heat fluxes over a tropical forest in eastern Amazonia," *Ecological Applications* **14**(4 Suppl.), 22–32.

Rosing, M. T.; D. K. Bird; W. Glassley; F. Albarede; N. H. Sleep (2006). "The rise of continents: An essay on the geologic consequences of photosynthesis," *Paleogeography, Paleoclimatology, Paleoecology* **232**, 99–113.

Sandberg, A. (1985). *The Y-Chromosome* (in two parts). New York: Liss Press.

Schwartzman, D. W.; T. Volk (1989). "Biotic enhancement of weathering and the habitability of Earth," *Nature* **340**, 457–460.

Sorokhtin, O. G.; S. A. Ushakov (1991). *Global Evolution of the Earth*. Moscow: Izdatel'stvo MGU [in Russian].

TEEB (Online). "The economics of ecosystems and biodiversity." Available at *http://ec. europa.eu/environment/nature/biodiversity/economics/index_en.htm*

Thomas, C. D. (1990). "Fewer species," *Nature* **347**, 237.

UNEP (2002). Available at *http://www.unep.org/dewa/assessments/ecosystems/water/vital water/05.htm*

Vinogradov, A. P. (1959). *The Chemical Evolution of the Earth*. Moscow: U.S.S.R. Academy of Science [in Russian].

Vinogradov, A. P.; Iu. A. Surkov; B. M. Andreichikov *et al.* (1970). "The chemical composition of Venus' atmosphere," *Space Research* **4** [in Russian].

Watts, J. A. (1982). "The carbon dioxide questions: Data wampler," in W. C. Clark (Ed.), *Carbon Dioxide Review*. New York: Clarendon Press.

Whittaker, R. H.; G. E. Likens (1975). "The biosphere and man." In: H. Lieth and R. Whittaker (eds.), *Primary Productivity of the Biosphere*. Berlin: Springer-Verlag.

Zavarzin, G. A. (2001). "The making of the biosphere," *Priroda* **11**, 988–1001 [in Russian].

Part IV

Sustainable development: Between complacency and reality

11

The basis of sustainability in nature and in civilization

Sustainable development: history of the term—"Development that meets the needs of the present without compromising the ability of future generations to meet their own needs"—Differences in the understanding and interpretation of development, growth and sustainability—Development without growth in living nature—Competition in the biota and in society—Genetic and extra-genetic mechanisms of inheriting patterns of sustainability—The rate of civilization growth and the evolution of the biota—A word from concerned scientists.

At this point let us take a closer look at the central concept of this book, namely, sustainable development—one of the most often used terms of the past two decades. Is sustainable development the guiding thread that leads the way out of the dead-end in which modern civilization finds itself or just another catchphrase good for attracting a few media bytes? Does anyone actually know what it really means? Until the latter part of the 20th century humanity somehow managed to get along without this term, relying on centuries-old, well-tested experience, all manner of compromise and the pursuit of narrow, selfish interests by specific national or social groups. Thus, in the past, international relations were usually based on temporary agreements or alliances that could be easily terminated in the face of changing power dynamics on the political chessboard. In other words, what we had was a form of spontaneous sociopolitical development, and the associated global ecological sustainability was made possible by the small human population, along with a low level of technological "weaponry".

However, in the early 20th century the situation changed radically as man finally mastered heretofore unknown sources of energy and acquired the ability to impact the environment on an unprecedented scale. And so, if until that point socio-economic cataclysms, wars and revolutions caused harm mainly at the local level—although this sometimes meant the elimination or devastation of entire peoples or states—a full-scale nuclear confrontation now had the potential to eliminate all life on Earth. The latter point was convincingly demonstrated by the computer models of Russian researchers under the leadership of academician N. N. Moiseiev who worked out the so-called "nuclear winter" phenomenon (Moiseiev *et al.*, 1985). There was also a hidden side to the destructiveness of our technology as exemplified by a seemingly harmless innovation patented in 1928—a coolant called freon. It was used widely in household cooling equipment, but a half-century later this great "step forward" ended up creating the ominous ozone holes above the poles of the planet.

On the other hand, when it comes to public awareness, a key accomplishment of the past 30 years or so is the general consensus that ecological sustainability cannot be viewed in isolation from social and economic parameters. To a large extent this is attributable to the realization that, given today's technological capacity, the most typical manifestations of corporate egoism can ultimately harm the interests of the egoists themselves. From the energy crisis of 1973 to the spiraling cost of crude oil in the first decade of the 21st century, our addiction to fossil fuels has harmed not only the environment, but also the economy. And so notions of social conscience and planetary stewardship have been harmonized in the minds of millions with purely practical short-term considerations, contributing to the development of an unprecedented idea: only a global effort can turn things around. This has led to the deflation of the pompous rhetoric typifying the discourse of many leaders and reformers in the first half of the 20th century (on both sides of the Iron Curtain). Against this background, humanity heard the phrase sustainable development coming for the first time "from on high"—at the United Nations Conference on Environment and Development in Rio where an alternative to the environmentally destructive *status quo* was presented to the world.

It was neither a surprise nor mere chance that the idea of sustainable development originated with the economically successful countries. Having destroyed their own natural ecosystems some time ago, they were the first to realize the ecological consequences for the rest of the world of an attempt to emulate their bankrupt policies. However, an analogous term appeared even earlier in the practice of natural management. In the mid-20th century the phrase sustainable yield was used by scientists and officials managing the fisheries of Canada. The idea here was a system of exploiting fish resources without exhausting supplies so that the yearly catch corresponds to the reproductive capacity of a given fish population. Unfortunately, this did not prevent the collapse of the northern cod fisheries off the Banks of Newfoundland leading to a fishing ban in the 1990s. However, at issue here is the idea of sustainable development that precedes the 1992 Rio Summit and even goes back to the thinking of German foresters a century earlier. They also considered the exploitation of forests whereby the rate of logging did not exceed that of natural growth (i.e., tree regeneration was supposed to occur without losses). Such resource use can continue indefinitely given the maintenance of factors unrelated to natural management. Already in the 1980s—largely thanks to the Report of the Brundtland Commission—the term "sustainable development" had migrated from conceptions of local ecosystem management to those of global ecology and gradually entered the wider scientific vocabulary.

However, it would be an exaggeration to claim that even today (a quarter of a century later) the world community has a clear view of the key elements constituting sustainable development or a unified viewpoint regarding its practical implementation. In fact, even the first definition of the term in the Report of the Brundtland Commission already gave rise to differences in interpretation. Thus, we read about the need for development "without compromising the ability of future generations to meet their own needs." However, it is not immediately clear what is meant by the needs of future generations—those of modern populations in the developed world or in countries where abject poverty has been overcome but the Western standard of living is still a dream? And how do we interpret the notion of development which "does not endanger the natural systems that support life on Earth: the atmosphere, the waters, the soils, and the living beings" (U.N., 1987) if no mention is made of the criteria determining the degree of such a threat? After all, man will always exert some sort of negative influence on the biosphere or to put it succinctly in the words of F. Nietzsche, "the world is beautiful, but has a disease called man." The question is how to prevent this "disease" from disrupting nature beyond the threshold of the biosphere's compensatory capacity.

In short, too many of these and similar statements appear to be ambiguous, insufficiently thought-out in scientific terms, and, most importantly, devoid of a serious theoretical foundation. This methodological fault—which is probably unavoidable at the beginning of a process where a problem is initially being grasped and formulated—subsequently led to numerous contradictory and at times quite random interpretations of a concept as fundamental for the modern age as sustainable development. Particularly cacophonic is the discourse regarding the theoretical compatibility of sustainability and growth or the relationship between growth and development (the latter two being particularly prone to confusion).

In this connection, some maintain that sustainability and development are in contradiction with one another, which implies that something in this combination must be given up (Valiansky and Kaliuzhny, 2002). In response to this we can point out that, from a philosophical position, development is a particular instance of movement (just as movement can be a particular instance of development) (e.g., movement toward a civil society, movement toward social justice, etc.). And the sustainability of motion (movement) is a fundamental mathematical concept first introduced by Henri Poincaré and subsequently developed by Aleksei Liapunov. It presupposes the kind of motion that begins from a point in a kind of imaginary tube and never leaves the confines of this tube. In other words, here motion is identical to change while sustainability corresponds to invariability—the constancy of some sort of relationship or property of an object maintained under any changing circumstances from a concrete fixed class (Danilov-Danil'yan, 2003). Applied to the topic at hand, we can argue that the development of civilization, a social group, or an economic system at a given point in time can be viewed as sustainable if it maintains a critical invariable element—in particular, if one is talking about such essential properties of a system that determine its very survival, for example. For civilization as a whole such an invariable element would be the limit of pressure exerted on the environment—a threshold beyond which the adaptive capabilities of the biosphere are exhausted and irreversible biospheric degradation begins (more on this topic in Chapter 14).

As for another pair of concepts—growth and development—disagreements about their semantics are partially attributable to the polysemy of both terms. The two overlap at some points and diverge at others. The verb "develop" includes such meanings as evolve, perfect, expand, and grow; all these concepts being applicable to business among other areas. The verb "grow" denotes such concepts as spring up, develop to maturity, increase in size, pass into a condition, acquire increasing influence, cause growth, and evolve. This appears to provide a foundation for those authors who link sustainable development with growth—albeit slowed-down growth limited by available resources or staying within the assimilating capacity of natural ecosystems (WB, 1994). At any rate, the vast majority of researchers allow for the possibility of one form or another of economic growth within the framework of sustainable development. And yet, as the above semantic ambivalence of growth and development demonstrates, the differences between the meanings of the two words are equally important.

Thus, while growth is a process of predominantly quantitative changes (as in "increasing in size"), development has to do more with qualitative ones (as in "perfecting"). For example, the constantly increasing pressure exerted by civilization on the biosphere is an instance of unbridled quantitative growth that rolls over all limits and can have very dangerous consequences. On the other hand, the fine-tuning of socioeconomic systems, their harmonization with the ecological parameters of the environment, as well as the accompanying evolution of our mentality and worldview, would constitute the kind of qualitative change suggested by some aspects of the word "development". And this would easily link up with sustainability as a partner concept of development. In this respect the inner workings of the natural biota stand in stark contrast with the notion of continual growth characterizing the outdated mindset of the 20th century's corporate dynamics.

With respect to nature, the term "sustainability" is linked to the evolution of natural ecosystems and is based on development without growth (or very limited growth). A natural tropical forest or a tundra community are well-established eco-systems that have been evolving but not growing territorially—at least in terms of historical (human) time. Such qualitative development appears to have no limits, which is attested by the immense complexity of the biota. The stimulus for this phenomenon is the constant dialogue of the biota with its environment and the constant "search" for more effective mechanisms of environmental regulation and stabilization. This includes processes for dealing with externally generated disruptions and means of returning the system to a stable state when perturbed. Such re-stabilization processes can sometimes require tens of thousands, or even millions of years in the case of particularly strong or long disruptions, such as great glaciations, because the internal structure of the biota is radically reorganized through the evolution of the species. All of this takes place on the basis of com-petition among organisms and their communities which pass though the sieve of natural selection in order to be tested for their ecological viability (i.e., the capacity to participate in the maintenance of environmental parameters required for life).

It would appear that, historically speaking, the evolution of humanity has been based on competition among ethnic groups, cultures, and civilizations. And yet, in this case—in contrast to natural ecosystems—constant and accelerating growth appears to characterize the demographic, economic, and material realms, the latter often being equated with progress. Furthermore, while in the case of the biota, competition is a key condition for its long-term stability, something entirely the opposite applies to human civilization. Here competitive relationships among groups and subsystems are usually the key source of instability in the human community. And if we consider the way human beings interact with the environment, it becomes apparent that here too our species is very different from all others on the planet. While the other species in one way or another inevitably adapt their activities to the prevailing environmental conditions, humans are the only beings that seek to adapt the environment to their needs. And this inevitably leads to environmental distortion and destruction.

If we turn to the ways in which the mechanisms of sustainability are inherited, again the biota appears to be fundamentally different from human beings. While genetic memory plays a key role in the biota in this connection, humans rely on extra-genetic memory (i.e., culture). Here we must distinguish between the basis—consisting of spiritual, moral, and other values that have to do with a certain world-view—and a complex of practical knowledge and skills which includes the various technologies at our disposal. While the basis of culture changes extremely slowly, constituting the nucleus of a stable society, practical experience and knowledge have been growing at an increasingly rapid rate. And the latter process has had tremendous impact on the natural environment. This is where the key clash between the wild growth of civilization and the biota's fundamentally slower pace of evolution becomes most evident. After all, as we increase our technological might, as well as our economic and financial capital, we cannot bring about a corresponding increase in the productivity of the "natural capital". The latter is determined by very different processes, such as the amount of solar energy received by the Earth, the capacity of the plant biota to assimilate that energy, the pace of biochemical reactions in living organisms, etc.

Therefore, despite the conceptual confusion outlined above, undeniably a breakthrough occurred through the Rio Summit's warning that the global ecosystem is not inexhaustible, that ecological factors must be part of the economic equation, and that technical progress is by no means necessarily the same thing as social progress. This turnaround in public consciousness made it possible to attract the attention of politicians, business circles, scientists, and the cultural sphere. The same year an alarm was sounded for the world community by a large group of scientists from 70 countries—in total around 1,700 people, including 104 Nobel Prize laureates, who were part of the Union of Concerned Scientists (UCS):

"Human beings and the natural world are on a collision course [...]. The earth is finite. Its ability to absorb wastes and destructive effluent is finite. Its ability to provide food and energy is finite. Its ability to provide for growing numbers of people is finite. And we are fast approaching many of the earth's limits. Current economic practices that damage the environment, in both developed and under-developed nations, cannot be continued without the risk that vital global systems will be damaged beyond repair [...] No more than one or a few decades remain before the chance to avert the threats we now confront will be lost and the prospects for humanity immeasurably diminished" (UCS, 1992).

At last, in this concept of finiteness that constitutes the spirit of the above declaration, there was a sense that key differences do exist between growth and development. The recognition of the vital importance of sustainability was put forth at the right place and time as part of the first conscious attempt to find a way out of the dead-end. The very foundations of human civilization were being reconsidered, and a step was taken beyond the environmentalism of the 1960s, whose weak and strong points were considered in Chapter 1. At this point let us consider the implementation of these ideas in the specific and concrete national plans and programs for sustainable development worked out on the basis of the Rio Summit's proposals.

References

Danilov-Danil'yan, V.I. (2003). "Sustainable development: A theoretical mathematical analysis," *Economics and Mathematical Methods* **39**, 2 [in Russian].

Moiseiev, N. N.; V. V. Aleksandrov; and A. M. Tarko (1985). *Man and the Biosphere: Systemic Analysis and Experiments with Models*. Moscow: Nauka [in Russian].

UCS (1992). *World Scientists' Warning to Humanity*. Cambridge, MA: Union of Concerned Scientists. Available at *http://www.ucsusa.org/ucs/about/1992-world-scientists-warning-to-humanity.html*

U.N. (1987). *Our Common Future*. New York: United Nations. Available at *http://www.un-documents.net/ocf-02.htm#I*

Valiansky, S. I; and D. V. Kaliuzhny (2002). *The Third Path of Civilization or Can Russia Save the World?* Moscow: Algorithm [in Russian].

WB (1994). *Making Development Sustainable*. Washington, D.C.: World Bank.

12

The national colors of sustainable development

Sustainable development American style: growth, prosperity, and a dynamic economy—China's Agenda 21: economic growth, birthrate control, following the American example—The "anti-ecological" direction of Russian policies—Development and growth as usual: avoiding the issue.

In the decade and a half that has elapsed since the Rio Summit, over 100 states have followed UNCED's Agenda 21 model by publishing their own agendas and programs in order to proclaim their respective visions of what constitutes sustainable development, as well as to define concrete steps to be taken toward these ends. The question arises as to what extent these national agendas correspond to the spirit and letter of the Rio Summit, as well as the documents adopted at the conference.

Let us consider the beginnings of the U.S. strategy for sustainable development as it was outlined in a document entitled *Sustainable America: A New Consensus for the Prosperity, Opportunity and a Healthy Environment for the Future* (PCSD, 1996) developed by the President's Council on Sustainable Development. This organization was created by the Clinton administration and included representatives from government, corporations, and NGOs. In the three years of its existence (1993–1996) the council welcomed contributions by any interested parties. Even in the preface to *Sustainable America* it is noted that "American society has been having increasing difficulty reaching agreement about societal goals. This has been especially true for those issues that lie within the overlapping shadows of Americans' hopes for economic progress, environmental protection, and social equity" (PCSD, 1996). These words express the objective contradictions inevitably faced by society as it tries to establish strategic priorities for sustainable development. In fact, this is where we find the crux of the problem.

The document's vision statement illustrates the difficulty of resolving this conundrum:

"Our vision is of a life-sustaining Earth. We are committed to the achievement of a dignified, peaceful, and equitable existence. A sustainable United States will have a growing economy that provides equitable opportunities for satisfying livelihoods and a safe, healthy, high quality of life for current and future generations. Our nation will protect its environment, its natural resource base, and the functions and viability of natural systems on which all life depends" (PCSD, 1996).

There is a certain contradiction here between the notion of "a growing economy" and the "viability of natural systems" (see Chapter 11). How could a country with only 4% of its natural ecosystems intact—as was the case in the 1990s (WR, 1990–1991)—maintain "functions and viability" of something that is no longer there? Unfortunately, no answer to this question can be found in the Presidential Council's document.

Next comes a "we believe" statement that can be viewed as a summary of the American principles for sustainable development. It contains 16 points, and 7 of them in one way or another have to do with economic growth, as well as economic development and effectiveness. Point 1, for example, states: "To achieve our vision of sustainable development, some things must grow—jobs, productivity, wages, capital and savings, profits, information, knowledge, and education–and others—pollution, waste, and poverty—must not" (PCSD, 1996). Thus, the key to this vision is yet again economic growth and prosperity—the phenomena meant to do away with environmental pollution and ensure the recycling of waste. However, one may ask in what way this view of development constitutes anything truly novel and different from the developed world's previous "environmentalist" economics responsible for making the global ecological picture worse. An example of this deterioration is the fact that at the time when *Sustainable America* was being drawn up, the U.S.A. alone was contributing 4.4% of the world's yearly increases in CO_2 emissions.

Chapter 1 of *Sustainable America*, entitled "National goals toward sustainable development" does not deviate in any significant way from previous policies and electoral promises of former U.S. governments. In fact, if one were to take out the words "sustainable development" from this text, we would have the impression of trying to walk forward on a treadmill. And the intention of assuming "a leadership role in the development and implementation of global sustainable development policies, standards of conduct, and trade and foreign policies" (Point 9) indicates the familiar desire to lead in a unipolar world in the near future.

So much for America's attempt to deal with the ideas coming out of the Rio Summit. But what about China—the evolving economic giant of the modern world? In 1994 the People's Republic also developed a program for sustainable development known as China's Agenda 21. This document appeared in 1997 under the title *The People's*

Republic of China National Report on Sustainable Development. Although this document fits into the tradition of previous five-year plans from the socialist bloc, the perspective is more long-term. As Chapter 2 of the China Report states, "this programme, taking into consideration the overall development strategy and in view of the 'Ninth Five-Year Plan and Long-Term Objectives for the Year 2010,' puts forth a series of counter-measures and suggestions" (Section 3.1 of the China Report). And in some instances the document looks ahead all the way to the year 2020 or even beyond.

The basic strategy in China amounted to intensive economic growth (up to 9% per year), but environmental protection and demographic control were also taken into account. The document notes China's large population and weak infrastructure, pointing out that only relatively rapid economic growth can eliminate poverty, raise the standard of living, and achieve lasting security and stability. Indeed, China is aware of the overpopulation problem more than any other country on Earth, encompassing 0.11 ha of agricultural land per capita. By the time China's Agenda 21 was written, its tillable land area had shrunk by 360,000 ha in 10 years, while its harvests amounted to less than 400 kg per capita. This prompted the authors of the report to call for the further expansion of family planning methods begun in the 1980s, as well as demographic control and population composition. The idea was to lower population growth to 1.25% by the year 2000, which did not occur as the per capita proportion of agricultural land fell to 0.07 ha.

As in the case of other countries with centralized, planned economies, China had virtually ignored ecological issues for a long time. However, by the 1990s this was beginning to change, as is attested by China's Agenda 21. The document discusses the necessity to check pollution and reach a partial improvement of the ecological situation in large urban centers. Special sections of China's Agenda 21 are devoted to natural resource protection and a sparing approach to their utilization. The preservation of biodiversity, measures against desertification (a particularly significant problem in China), solid waste recycling, and atmospheric protection are further issues addressed by this document.

The report also adopts a global strategic approach along the lines of *Sustainable America*. Turning to the American vision of sustainable development—whatever the timeframe in question may be—the authors of China's Agenda 21 view it as a model for the rest of the developing world. In particular, an argument is made that the U.S.A. should use its wealth and technological capabilities to develop an effective domestic policy for achieving sustainable development, thereby demonstrating a more rational path toward progress. However, at this point one might ask how this "rational path toward progress" can be followed by countries that lack the necessary economic and technological prerequisites. Will they end up repeating the American example by exhausting not only their own resources, but also those of other states? And here we are back to where we started.

Thus, China chose essentially the same strategy as the U.S.A., taking into account local variables and considering even greater rates of economic growth. The latter, ironically, appeared to the authors of China's Agenda 21 as part of the solution to the environmental protection problem—an area where the developing giant was only beginning to take its first cautious steps. However, even in terms of economic growth a long time will pass before China is in a position to approach the United States because the Chinese GDP is still far below its American counterpart.

The next country to consider with respect to this issue is another geographic heavy-weight—Russia. Here there has been no official state program for a transition to sustainable development; however, Russia's sheer territorial might is associated with a natural potential that will undoubtedly play a key role in determining the stability of the global ecosystem. Therefore, the rest of the world must take the Eurasian giant into account as it works out its ecological strategies.

As in China's case, Russia's centrally planned centralized economy of the Soviet period had left virtually no room for environmental protection. So, here too, ecological issues were being considered for the first time, although there were and still are more than enough causes for alarm all over Russia's territory. For example, Russia has 11 million km^2 of well-preserved natural ecosystems (around 65% of the entire country or 22% of the world). However, two-thirds of the population lives in ecologically disrupted areas that occupy 15% of the country's territory. The latter figure corresponds to 2.5 million km^2—which is more than the U.K., France, Germany, and Scandinavia put together. Here we find almost all the major urban centers of Russia—around 150 cities—where allowable concentrations of toxic substances in the air and drinking water are constantly exceeded. And although atmospheric pollution was gradually decreasing in the 1990s, yearly growth has been increasing since 2000 (see Table 12.1). Russia's urban and suburban dumps (which are not very different from ordinary waste disposal sites) have accumulated over 110 billion tons of solid industrial and household waste. The latter poisons ground water, as well as surface water bodies, and fills the air with dangerous dioxins (Danilov-Danil'yan, 2003). Nothing of the sort would be encountered in any developed country of the world, but for most of Russia's population a poor ecological background is part of daily life.

And yet, today tens of millions of people in Russia appear to view sustainable development as a remote problem having no direct connection to their daily concerns and future prospects. This may be the case, as demonstrated by researchers from the Club of Rome, because the notion of tomorrow usually does not extend beyond a few weeks ahead. Given the difficulties of life in Russia where much of the population is consumed by the struggle for basic needs or even survival, it is difficult to make people focus on anything beyond the daily grind (cf. Losev, 2001). Therefore, it is no surprise that the media have typically followed this mass psychology, bombarding Russia with "hot" news (i.e., sensationalist material that offers virtually no food for thought regarding the ultimate fate of the country or the world as a whole).

As for Russia's current political leaders, most of them appear to think that in this time of restructuring, questions of sustainable development are nowhere near the top of Russia's agenda. The logic of this thinking suggests that current difficulties must be overcome first, and only afterwards can one start considering the transition to anything like sustainability. This is coupled with the indifference of the Russian

Table 12.1. The dynamics of atmospheric pollution from standard sources in thousands of tonnes (*State Report on the Environment of the Russian Federation*, 2000, 2003, 2004).

	Years						
	1996	*1999*	*2000*	*2001*	*2002*	*2003*	*2004*
Russian Federation	20,274	18,540	18,820	19,124	19,481	19,829	20,491
Industry	16,661	14,704	15,222	15,492	15,842	15,875	16,733
Electrical energy	4,748	3,935	3,857	3,656	3,353	3,447	3,258
Non-ferrous metallurgy	3,598	3,312	3,477	3,405	3,297	3,262	3,287
Oil industry	1,309	1,329	1,619	2,119	3,113	3,227	4,195
Iron and steel industry	2,535	2,330	2,396	2,268	2,223	2,178	2,203
Coal industry	596	560	604	786	819	764	757
Oil-processing industry	850	748	736	679	621	594	581
Gas industry	542	456	501	476	537	591	651
Construction materials industry	528	417	441	455	434	448	473
Chemical and petrochemical industry	413	415	427	437	428	403	408
Machinery production and metalworking	602	454	433	433	370	356	340
Wood processing and pulp/paper industry	434	367	379	372	332	309	304
Food industry	250	198	182	168	163	155	143
Light industry	64	51	45	44	41	34	27
Other industries	504	368	351	405	313	296	107
Housing and utilities infrastructure	658	943	981	999	1058	1078	991
Agriculture		111	121	133	126	127	119
Transportation	2,370	2,394	2,062	2,055	2,005	2,175	2,137
Incl. communal pipeline transport	2,024	2,111	1,797	1,787	1,720	1,890	1,826
Other areas of the economy		388	435	444	450	576	511

electorate toward ecological questions with seemingly no direct bearing on one's area or village, and so political platforms and programs ignore the issue altogether.

However, this was not the case when Gorbachev's *perestroika* was in full swing in the late 1980s and during the early Yeltsin period in the first half of the 1990s. In 1989 ecological issues were very much a topic of public and political awareness as the country discovered for the first time to what extent the Soviet system had been polluting the environment and denying it in the usual demagogic manner. By the year 2000, in the wake of the chaos, mismanagement, and corruption under the Yeltsin administration, ecology was long forgotten (Losev, 2001). People were far more concerned with organized crime and unpaid salaries than acid rain or oil spills. To this day Russia remains far behind when it comes to public environmental concern. Only a handful of specialists and professionals discuss such questions at isolated conferences, about which hardly anyone knows or hears.

Exemplifying this trend is the fate of ecological institutions in the past 20 years. In the early post-*perestroika* period there was a ministry of the environment. However, it was soon demoted to the status of a State Committee for Ecology and Natural Resource Protection. When Vladimir Putin came to power in the year 2000, the committee was eliminated, and its jurisdiction was transferred to a branch of the Ministry for Natural Resources. In other words, the Soviet patterns were being resurrected: the very institution responsible for pollution was charged with doing the cleanup and overseeing its own actions. The results were predictable as ecological standards for economic activity diminished dramatically. Big and medium-sized businesses have received significantly more access to natural resources. The best natural areas around big cities are undergoing major development as they turn into elite cottage country and secondary residence communities while supporters of nature reserves constantly have to fight for their survival against the development plans of local officials and their business friends.

There have been attempts to create the impression that Russia's leadership does care about the environment. Thus, in 1994 President Boris Yeltsin issued a decree on "The State Strategy for the Development of Environmental Protection and Sustainable Development in the Russian Federation." Two years later another presidential decree on sustainable development appeared. However, the Yeltsin administration failed to adopt a key document (in the eyes of most countries) entitled *Strategy for Sustainable Development* and prepared in the second half of the 1990s by a large group of specialists, *Duma* deputies, and community representatives. Instead various interim or "palliative" documents have appeared here and there. For example, the Russian government approved the Strategy for the Socio-Economic Development of the Russian Federation up to 2010 (the so-called "G. Gref Program"). In 2002 the *Ecological Doctrine of Russia* appeared as a smokescreen to conceal the widespread public indignation in the wake of a decision to import radioactive waste into the country. Needless to say, none of these documents can replace a full-scale program for true sustainable development.

As if to illustrate the expression "too many cooks spoil the broth", there are around 20 disjointed Russian ministries and jurisdictions that deal with ecological problems. The absence of a unified environmental protection agency amounts to a virtual state of paralysis in this matter (Fomin, 2005). As a result, Russia lags dangerously behind most other countries as it drifts in an anti-ecological direction, which has negative consequences for its general economic development. This is particularly well exemplified by the growth of the unit cost of environmental resources (the environmental capacity of production) in the past two decades, along with the increase in the percentage of economic branches that exploit and pollute nature. Furthermore, to this day the end-product unit cost of natural resources is still extremely high in Russia. This points to the country's inefficient economy that still operates on the assumption of a limitless resource base. Therefore, the energy capacity of final production is two to three times higher in Russia than in the developed countries while the use of forest products for the production of 1 ton of paper exceeds that of developed countries by four to six times (Korochkin, 2002).

The above trends make it perfectly plain that the Russian economy will soon consist mainly of the natural resource sector with a small share of machine building and high-tech production (mainly in connection with the arms industry). However, tomorrow's economics evolves from today's. And if the current system already embodies an anti-ecological core, no amount of environmental protection will save Russia from itself in the future.

We have examined three key economic systems—the United States, China, and Russia—with regard to the question of making the transition to sustainable development in the wake of the Rio Summit. These countries poignantly embody the features of the developed and the developing worlds, as well as those of the former socialist bloc which are representatiive of countries with so-called transitional economies. The situation described above hardly corresponds to the spirit and letter of the Rio Summit and the Brundtland Commission—a conclusion that can apply just as easily to the majority of the developed and developing nations. In fact, a survey of the official documents adopted after the Rio Summit—Anon. (1997); Canadian Government (1997); FMENCNS (1997), and others—indicates that in virtually the entire developed world, sustainable development is still viewed in terms of a strategy familiar to us from the past three decades. The environment is taken into account to be sure (especially with respect to waste and pollution). However, no serious limitations are placed on economic growth. No rigid ecological boundaries are set for the sphere of production—which in the final analysis suggests that references to a global collision between humanity and its environment are merely empty words. As we continue to produce more TV sets, cars, and airplanes, very few political leaders are willing to admit that the economic pufferfish can puff itself up only so much before it bursts.

References

Anon. (1997). *From Environmental Protection to Sustainable Development.* Stockholm: Gotab, 1997.

Canadian Government (1997). *Building Momentum: Sustainable Development in Canada.* Ottawa: Canadian Government.

Danilov-Danil'yan, V. I. (2006). "The ecological significance of energy conservation." In: *Russia's Energetics: Problems and Prospects: Proceedings of the Scientific Session of RAN*. Moscow: Nauka, pp. 196–207.

FMENCNS (1997). *Towards Sustainable Development in Germany*. Bonn: Federal Ministry for the Environment Nature Conservation and Nuclear Safety.

Fomin, S. A. (2005). "The main executive government agencies of Russia in the area of ecological management in 2005," *Russia in the Environment*. Moscow: Modus-K Eterna [in Russian].

Korochkin, E. F. (2002). *Ecology and Sustainable Development in Russia*. Moscow: Ministry of Natural Resources of the Russian Federation. Available at *http://www.mnr.gov.ru/files/part/7239_korochkin.doc* [in Russian].

Losev, K. S. (2001). *Ecological Problems and Prospects for Sustainable Development in Russia in the 21st Century*. Moscow: Kosmosinform [in Russian].

PCSD (1996). *Sustainable America: A New Consensus for the Prosperity, Opportunity and a Healthy Environment for the Future*. Washington, D.C.: The President's Council on Sustainable Development. Available at *http://clinton2.nara.gov/PCSD/Publications/TF_Reports/amer-top.html*

PRC (1997). *The People's Republic of China National Report on Sustainable Development*. Beijing: People's Republic of China. Available at *http://www.acca21.org.cn/nrport.html*

WR (1990–1991). *World Resources Report, 1990–1991*. Washington, D.C.: World Resources. Available at *http://www.wristore.com/worres19clim.html*

13

Co-evolution of nature and society: Fact or fiction?

Is there any need for new terms: "co-evolution" or "sustainable development"—The rate of the innovation process and speed of the formation of "natural technologies"—A question without an answer: the limits of biospheric stability—Evolution toward humanity or development toward the biosphere—The reality behind the term "noosphere".

It seems evident that the transition to sustainable development in order to prevent a biospheric catastrophe is a goal that surpasses the scale of all objectives faced by humanity at any time or in any place. Since the 1960s—when global ecology first became the subject of careful scrutiny on the part of the international scientific community—we have certainly made major advances when it comes to grasping the issues on an intellectual level. However, when it comes to concrete action, our achievements are puny compared with the extent of the increasing harm experienced by the environment in the same period. We cannot even claim that we are standing still, but instead there has been a constant retreat on all ecological fronts. The gap between good intentions and reality has elicited various reactions, such as the idea of human and biospheric co-evolution. To quote N. N. Moiseiev: "At the Rio Summit an attempt was made to formulate a certain common stance or scheme of behavior for the planetary community. This became known as sustainable development which can be equated with the co-evolution of humanity and the biosphere" (Moiseiev, 1997). However, given the worldwide use of the term "sustainable development", what is the point of resorting to the concept of co-evolution? And, furthermore, are the two as identical as Moiseiev views them?

In fact, Moiseiev is not the only one to draw an equal sign between these notions—as various Russian publications show(e.g., Rodin, 1991; Karpinskaia *et al.*, 1995). Originally the term "co-evolution" meant merely mutual adaptation or coordinated changes in various species in the course of biological evolution. However, some researchers began to apply this word to a wider range of phenomena related to the development of any interacting subsystems or elements within one overall system. In this sense, co-evolution has to do with the mutual adaptation of evolving systems whereby changes in one system do not harm the functioning of the other.

In 1975 E. Odum outlined nine types of interaction among biological populations, viewing them more or less justifiably as varieties of co-evolution (Odum, 1983). However, applying this concept to the study of co-evolution between nature and society is a task so complex and unique that a very significant effort is required. In this connection let us turn to Moiseiev's idea that humanity is part of the biosphere, which is a hardly disputable concept. One question that arises here is the asymmetrical situation created by the co-evolution of a part (humanity) and the whole (biosphere). However, we can overlook this problem for the time being and consider exactly what is meant by the evolution of the biosphere, on the one hand, and the evolution of human society, on the other.

A classical systemic understanding of the environment leaves room for only one satisfactory definition of the biosphere: a system that encompasses the biota (the totality of all living organisms including man) and its surroundings (the totality of objects that either are influenced by the biota, or themselves act upon the biota). However, in our case, we are not focused on any influences in general, but rather specifically those that can affect the fate of civilization and survival of the human species. In such a context we can talk about acceptable/unacceptable or desirable/undesirable changes in the biosphere. As for the evolution of the biosphere, the role played by living organisms in the formation of the ocean, the atmosphere, and soil and rocks suggests the primacy of the biota. The latter evolves through speciation, and the appearance or disappearance of any given species—given the systemic nature of the biota—leads to an inevitable wave of changes in natural ecosystems where the organisms in question are "inscribed". The rate of this process is determined by the duration of a given species' existence (on average, 3.5 million years) and the period of its formation (of the order of 10,000 years). There are reasons to believe that these figures have remained constant for at least a few hundred million years (Danilov-Danil'yan, 1998).

The evolution of human society, on the other hand, follows completely different patterns. Genetic parameters remain relatively invariable within *Homo sapiens*. What does change is social structures, social consciousness, material and spiritual culture, as well as the productive and scientific–technological potential. The latter has determined humanity's impact on the biosphere, involving the kind of innovation process that is somewhat reminiscent of speciation in the biota. For this reason it is informative to compare the relative rates of species formation and technological progress. Material production appears similar to the biota in that it is characterized by systemic, spontaneously formed organization. Any innovation (the appearance of a new technological element in the sphere of production and management) leads to a wave of other innovations in the corresponding technological niche. However, while the rate of biological evolution has remained virtually unchanged for millions of years, the rate of technological change has been constantly growing. For example, by the end of the 20th century, the duration of the innovation cycle in the realm of advanced technology was only 10 years on average.

Now two numbers can be juxtaposed—10 years for the creation of new industrial technologies and 10^4 years required for the appearance of new "natural technologies" (i.e., new species). Given this difference, amounting to three orders of magnitude, is it justifiable to bring up the possibility of something as cooperative and reciprocal as the co-evolution of humanity and nature? And if so, then how do we imagine the co-evolutionary changes in the biosphere in response to human economic and technological innovations? Are we conjecturing the appearance of species whose evolution can keep up with the rate of anthropogenic change (e.g., bacteria capable of decomposing polyethylene or transforming mountains of discarded aluminum cans into bauxites and nephelite)? Could humanity perhaps create new species in a test tube or modify the genetic makeup of existing ones in order to speed up the speciation process in the biota and increase its co-evolutionary capabilities? The future keeps surprising us, so the above cannot be ruled out. However, even if any of these "Wellsian" scenarios were possible, the missing element in this co-evolutionary scheme would be reciprocal action (i.e., the natural world would be adapted to the human one but not the reverse). Unfortunately, the one-way basis of our relationship with the environment is attested to by thousands of years of history and prehistory—from the taming of fire to the invention of agriculture, etc.

Clearly, the biosphere ensures the maintenance of environmental parameters essential for life and, like any highly organized system, it is characterized by a stability threshold. However, once we start interfering in the evolutionary process, how can we determine which of these parameters is essential, and what are the limits of allowable change? Can we guess which of the biosphere's subsystems (biological species, communities of organisms) can survive violation of the above-mentioned stability threshold and which ones cannot? As in the case of the butterfly effect from chaos theory, the disappearance or modification of any biological species through human action is an event that can disorganize the biosphere and unleash a wave of consequences fatal to other species or communities. Could this lead to a "tsunami" that will engulf the biosphere as a whole? Not according to those who place their money on a technological solution to the problems created by civilization. They argue that whenever human science and technology destabilize the environment, they can turn it around and restabilize it.

Indeed, there are some examples of successful attempts to patch up man-made holes in the biospheric balance. The International Birding and Research Center—Eilat (IBRCE) is one of them. In recent times seasonal shelter has been given to millions of birds, including raptors, passerines, and water birds (a total of several dozen species) which migrate yearly back and forth between Eurasia and Africa. The center is located on the Israeli coast of the Red Sea—in a depression of the Gulf of Aqaba. This is a key area within one of the world's three migration corridors where birds arrive exhausted after flying for many days over the Sahel, Sahara, and the Red Sea. They have no other rest-and-recovery area, since the Arabian desert stretches to the east for thousands of kilometers while the Dead Sea and the Negev Desert lie straight ahead where no food can be found.

In the 1960s, an oil pipeline and tanker complex were built here, which was followed by the construction of a resort. Fortunately, however, it became clear that this activity had deprived migrating birds of their only available harbor, and there was still enough time to rectify the situation. In 1993 a famous Israeli ornithologist Reuven Yosef started an initiative—which involved a battle with the local hotel industry. A covered municipal landfill site was used to create one of the world's biggest ornithological stations. Freshwater reservoirs were dug, the area was fenced in to keep outsiders out, and, most importantly, the feathered guests were provided with food. Thus, the Eilat center became not only a location for observing and ringing birds, but also a kind of bird hotel or migration top-up station. Its operation has already required over two million dollars, and it can be safely claimed that the enthusiastic work of the center's small staff has affected the biological balance of huge territories—from northern Scotland to the steppes of the Black Sea region (Yosef, 1996, 2002).

However, this success story is the exception rather than the norm. It merely means that humans can sometimes prop up the collapsing ecological building, but other parts of the structure are bound to sag in their turn. There is not a single instance of environmental protection initiatives by technological means whereby a local improvement was not achieved at the expense of deterioration in the overall ecological balance. Simply put, the amount of energy required for such objectives is so high that the benefits are outweighed by the losses. And even if we imagine that a breakthrough in this field may be possible in some distant future, that stage is so remote and nebulous that we simply do not have the time before the current ecological crisis turns into a biospheric catastrophe, putting an end to all our ambitions.

Therefore, no matter how we approach the issue—ecologically or technologically—there are no convincing reasons to assume the viability of co-evolution between the biosphere and society. And so, since the biosphere will clearly not evolve toward us, we might consider something like the opposite. This would involve the gradual weakening of anthropogenic pressure stemming from a reconsideration of our values (i.e., the basis underlying the functioning of modern civilization). This is (or should be) the core of true sustainable development—the only chance that humanity has to prevent the irreversible.

While co-evolution is a relatively recent term, the associated *noosphere* goes back to the 1920s. It comes from the Greek *noos* (mind) and refers to a global consciousness or a "thinking planetary shell" of sorts. The word was introduced by two French Bergsonians—the mathematician Edouard le Roy (1870–1954) and the anthropologist Pierre Teilhard de Chardin (1881–1955). We must also mention the important contribution to the noosphere concept made by Vladimir Vernadsky. In 1922 he was invited by the Sorbonne to give a series of lectures on geochemistry—a new science that Vernadsky was actively developing. The ideas brought by Vernadsky to Paris included more than just geochemistry, notably the recently formulated notion of the biosphere as a planetary layer that unified all the non-living and living substances with the latter having primacy in the shaping of the Earth's surface. With respect to these processes, Vernadsky accorded a special place to the human species that had already become an independent geological force.

Neither Le Roy nor Teilhard had any data regarding the influence of humanity on the biosphere. In fact, until their meeting with Vernadsky, they knew nothing of the biosphere. Vernadsky spent four years in Paris, lecturing at the Collège de France to an audience that included Le Roy and Teilhard. He also spoke at the Henri Bergson seminar although, as we know from other participants, Vernadsky did not personally meet the Nobel Prize winner himself—whom he greatly revered. The result was mutual intellectual stimulation. Le Roy and Teilhard were given the chance to place their noosphere concept on the solid footing of the natural sciences. Vernadsky, for his part, developed the idea of his French colleagues in order to posit the noosphere as a stage in biospheric evolution, thereby making a major contribution to this field of inquiry (Nazarov). Here is what Vernadsky wrote in the late 1930s: "Humanity as a whole is turning into a major geological force. Our thoughts and endeavors are gradually bringing us to the question of reconstructing the biosphere in the interests of a freethinking human race as a whole [. . .] Under the influence of scientific thinking and human labor the biosphere is moving into a new state—the noosphere" (Vernadsky, 1993, p. 305; 1988b, p. 27).

However, as noted before, this was a time of social optimism and in many instances—the belief in the bright future of mankind (especially in the Soviet Union). Despite the shadow of fascism hanging over Europe and the horrible suffering caused by the communist experiment in the U.S.S.R., Vernadsky associated his optimism for humanity with the boundless potential of human reason and the very essence of modern civilization:

"The civilization of cultured humanity—the organizational form of a new geological force that has appeared in the biosphere—cannot be cut short or destroyed because it is a great natural phenomenon corresponding to the historically, or rather geologically shaped structure of the biosphere. Civilization makes up the noosphere and is linked by all its roots with this planetary layer, for which there is no precedent or anything even remotely comparable in human history" (Vernadsky, 1988a, p. 46).

Neither Vernadsky, nor Le Roy, nor Teilhard had any doubt that human economic and technological influence on the biosphere was overall a positive phenomenon although they did think that this spontaneous process had to be channeled in some rational manner. Arguing for conscious purposefulness of this process, Vernadsky saw no danger in the scale of what we were doing. All his faith was placed in science that opened up heretofore unheard-of possibilities and prospects: "The coordination of human activities and ideas all over the world cannot be stopped. As far as the near future is concerned, scientists have the monumental task of directing the organization of the noosphere in a conscious manner. There is no turning back since the natural and spontaneous expansion of knowledge pulls us all in this direction" (Vernadsky, 1988b, p. 50).

Needless to say, the many decades that have passed since the death of the three men who developed the concept of the noosphere have greatly changed the world around us. And yet, however misguided some of Vernadsky's optimism may have been, much of what he foresaw has come true. We are indeed witnessing the process of globalization and a unified informational space (internet, television, cellphones, satellites, etc.), corresponding to Verndasky's vision of "instant thought transmission and simultaneous discussion of ideas all over the planet" (Vernadsky, 1989, p. 136). Furthermore, we owe a debt to Vernadsky who looked beyond the chaos of confrontation that characterized the world of the 1930s and called for a stop to the spontaneous evolution of humanity on the basis of our responsibility for the future of the biosphere. Perhaps a key change in our understanding of the biosphere compared with that of Vernadsky is the recent humbling of scientific ambition. The purportedly boundless ability of the human species to manipulate its environment has turned out to be far less than what it seemed in the first half of the 20th century. And most importantly, we now know that our former paternalistic attitude and our current scientific–technological potential are no match for the global processes ensuring biospheric stability.

It is in this connection that Vernadsky's idea about the transition of the biosphere into a qualitatively different state—the noosphere—elicits considerable doubt. As we know on the basis of biotic regulation theory, the shift of the biosphere into any other state is only possible through a disruption that pushes the system past a certain threshold. Disruptions of such magnitude have taken place during periods of great glaciations after which the return to a new level of stability was made possible by the radical reorganization of the biota's structure—a process stretching at times over millions of years. If human activity reproduces a disruption of this type, and even if the results are not irreversible, the time required for biospheric recovery would be too great to make any difference for us. When it is all over, neither our species nor many others are likely to have a place in the radically new biosphere. Obviously this was not what the authors of the noosphere concept had in mind.

References

Danilov-Danil'yan, V. I. (1998). "Is the co-evolution of nature and society possible?" *Questions of Philosophy* **8**, 15–25 [in Russian].

Karpinskaia, R. S.; I. K. Liseev; and A. P. Ogurtsov (1995). *The Philosophy of Nature: Co-Evolutionary Strategy*. Moscow: Interpraks [in Russian].

Moiseiev, N. N. (1997). "Co-evolution of nature and society: The way of noospherogenesis," *Ecology and Life* **2/3**. Available at *http://www.igrunov.ru/vin/vchk-vin-discipl/ecology/authors/Moiseiev/noosferogenez.html* [in Russian].

Nazarov, A. G. (2001). "In search of the noospheric reality." Available at *http://www.vernadsky.ru/Noosfera/Bul_11/11-8.htm* [in Russian].

Odum E. P. (1983). *Basic Ecology*. Philadelphia: Saunders.

Rodin (1991). *Idea of Co-Evolution*. Novosibirsk: Nauka [in Russian].

Vernadsky, V. I. (1988a). *The Philosophical Ideas of a Naturalist*. Moscow: Nauka [in Russian].

Vernadsky, V. I. (1988b). *Scientific Thought as a Planetary Phenomenon*. Oxford, U.K.: Routledge. Available at *http://vernadsky.lib.ru/e-texts/archive/thought.html*

Vernadsky, V. I. (1993). "A few words on the noosphere," *Russian Cosmism: An Anthology of Philosophical Thought*. Moscow: Pedagogika Press [in Russian].

Yosef, R. (1996). "Eilat, Israel: Avian crossroads of the Old World," *Living Bird* **15**, 22–29.

Yosef, R. (2002). *Pollution in a Promised Land: An Environmental History of Israel*. Berkeley and Los Angeles: University of California Press..

Part V

On the scale of a scientific approach

14

Sustainable development in relation to the carrying capacity of the biosphere

Warning number one for humanity—The law of energy stream distribution in biota—Key players on the stage of life—The energy corridor for the existence of large animals and man—The carrying capacity of the biosphere as an integral of the biospheric disruption limit—The 1% energy quota of civilization—The geographic equivalent of the biospheric disruption threshold—Once more about the possibility of technogenic environmental regulation

Given our relative well-being and economic growth as well as the impressive success of scientific and technical progress of the 1960s and 1970s, why did the question of sustainable development suddenly arise? Because the most insightful scientists realized that humanity had approached a certain critical limit, encountering external boundaries in its development. Admittedly, at first, such limitations were perceived as being essentially based on natural resources—a position that predominated at the time in the Club of Rome reports. However, ecologists following the most consistent lines of reasoning concluded that such boundaries were determined not so much by subsurface resources or accessible energy sources, as they were by the biospheric potential for the neutralization of growing anthropogenic disruption. The latter is inevitably bound up with the attainment of a critical threshold where this potential is exhausted—something which we are essentially experiencing today. In light of this, the biosphere—unable to cope with anthropogenic pressure—is reaching a stage of degradation (perhaps reversible at first) that will presumably continue until its cause disappears. That cause is human civilization that has been unable to harmonize its development with the adaptive capacity of the natural environment. And all this could happen much earlier than the explosion of a real crisis related to one of our vital resources.

However, while human economic activity constantly destabilizes existing natural interrelationships, natural biota, on the other hand, from the moment of its appearance has been constantly facilitating the maintenance of environmental sustainability and thereby ensuring the conditions for its own existence. In the early stages of life's evolution more than 2 billion years ago, this work was carried out by unicellular prokaryotes that formed the basis for the modern biogeochemical machine (Zavarzin, 2000). Later this job was taken on by multicellular organisms as well—mainly plants and fungi—which together constitute most of the biomass, fill the atmosphere with oxygen, consume excess CO_2, and take part in the formation of sedimentary rock. Zooplankton and phytoplankton—minute animal and plant organisms living in the ocean's surface layer—provide the leading role in the World Ocean's stabilization of the planet's environment. Now that civilization has destroyed more than 60% of natural ecosystems on land, it is the oceanic depths—with their (so far) only mildly disrupted biota—that act as the main channel or drain for the elimination of excess anthropogenic carbon from the atmosphere. However, even the World Ocean is having trouble coping with the various forms of the growing anthropogenic strain.

Thus, according to existing estimates (Houghton *et al.*, 1996; Gorshkov *et al.*, 2000), the ecosystems and waters of the World Ocean today absorb more than half of the atmospheric fossil fuel carbon emitted during burning, while the rest accumulates in the atmosphere. The ocean's ecosystems also absorb approximately 2/3 of the so-called "excess" carbon that forms in areas of dry land disrupted by economic activity. The remaining 1/3 is absorbed by ecosystems still remaining on land—primarily the forests and wetlands of Russia and Canada (cf. Chapter 15). It is clearly evident that when the integrity of a key biogen's cycle is violated, the biogen gradually accumulates in the atmosphere. And this, among many other facts, is the primary element indisputably indicating that the allowable limits of human impact on the biosphere have already been exceeded. There is no doubt that this impact has proceeded beyond the biosphere's carrying capacity.

The notion of the biosphere's carrying capacity as a key indicator of human material activity is being introduced for the first time on these pages. Although this concept occupies a fundamental position with respect to sustainable development, offering a tool for essential quantitative analysis, it is not interpreted consistently in scientific circles. In this connection, the theory of biotic regulation makes it possible to consider this notion from a fundamentally new and strictly scientific angle.

First, as with any other species on Earth, humanity exists within a certain "energy corridor". This corridor is characterized by the maximal share of the general energy stream in biota that humanity can use for its needs without the risk of disrupting the environment. Naturally, at issue is energy that has already been transformed by phytoplankton and land plants through the process of photo-synthesis—energy that these life forms store in the form of organic matter called *primary production*. The yearly yield of this organic matter created on any given territory is known as *gross primary production*. However, approximately 15% to 70% of the energy stocked up by plants is expended for their own growth, respiration, and proliferation (Ricklefs and Miller, 2000). Therefore, the subsequent cycle includes only the remaining percentage of the energy that is used by consumer-organisms at the next trophic levels. This energy constitutes so-called *net primary production*—a concept that was introduced in previous chapters. The yearly autumnal leaf fall as well as falling dry twigs and fruit at moderate latitudes are typical examples of net primary production streams.

However, this is merely the external manifestation of the net primary production stream. Its essence is the transfer of energy contained in organic plant matter from one group of organisms to another or from one trophic level to the next. The sum total of such shifts can reach four or five, and even six in some cases. Measurements taken in numerous different undisturbed natural ecosystems have demonstrated a clear pattern in the distribution of this energy stream depending on the size of organisms. This pattern applies to very different natural communities. In this manner, it has been established that about 90% of net primary production in natural ecosystems is consumed by bacteria and fungi which also play a leading role in environmental regulation and stabilization. Approximately 10% of net primary production is used by invertebrates, such as arthropods, worms, mollusks, etc. Less than 1% of the energy circulating in biota is consumed by the vertebrates—mainly large mammals that are responsible for a finer level of functional fine-tuning of natural communities (Figure 14.1). These features are indisputable and manifest a high level of stability: that is, they maintain (or at least maintained until recently) their values within a very narrow range of possible fluctuation over tens of millions of years (Gorshkov, 1981; Gorshkov *et al.*, 2000).

As was stated in Chapter 10, the main goal of biota in global and local terms is maintaining a high integrity level of the substance cycle, without which the environment would quickly degrade to a state unsuitable for life. This is made possible by a huge number of uncorrelated, autonomous individuals or organisms sharing a weakly intra-correlated structure constituting the framework of every local ecosystem. Only in this manner—on the basis of the statistical law of large numbers—is it possible to minimize random fluctuations in the flow of synthesis and decomposition of organic matter. In other words, in order for synthesis and decomposition to be correlated with each other, they must be carried out by many independent operational units. This is confirmed by theoretical studies, as well as experimental data.

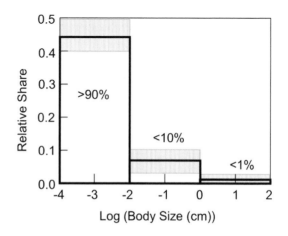

Figure 14.1. Distribution of the relative share of consumption of net primary production of terrestrial ecosystems of different body sizes. Percentages indicate contributions from each bar in the histogram (based on Gorshkov *et al.*, 2000, fig. 3.7.1).

As we stated above, 90% of photosynthesized organic matter (95% if plant respiration is taken into account) is decomposed by bacteria and saprophytic fungi. However, while every bacterial cell is an independent organism, the fungi consist of thread-like formations (hypha) a few microns thick and constitute a rather weakly intra-correlated multicellular structure which can be seen by the naked eye or under a magnifying glass. The vegetative body (the fungus mycelium) looks like a spider web, a velvety tarnish, or an extremely thin film. At the same time, the length of fungus filaments on the forest floor can reach 35 km per gram.

Something similar occurs in plants as well. They also constitute a multi-cellular structure consisting of weakly inter-correlated parts (modules) which are characterized by widely differing lifespans. Thus, tree leaves in temperate latitudes can survive for only one season, which cannot be said for roots and trunks. Therefore, when one part of a plant dies (e.g., when branches are gnawed by herbivores), the whole organism does not die, but instead this stimulates the development of the plant's other parts. This does not take place in highly intra-correlated animal organisms. In other words, strictly speaking, trees are not whole organisms but rather agglomerations of weakly interconnected modules—roots, branches, flowers, and leaves— which are only partly inter-correlated by the overall woody structure and a vascular system responsible for transporting water and nutrients from the soil. At the same time, the leaves on one tree can compete among themselves for light and biogens. And so, in the case of plants, fungi, and bacteria, the decrease in fluctuations that characterizes synthesis and destruction in a biotic community is achieved in essentially the same way. Ecological complications arise only when large animals disrupt the "calm" of an ecosystem.

Let us begin by considering the consumers of the first order—lagomorphs, ungulates, primates, and other animals who feed on plant biomass. It has been calculated that the metabolic capacity of birds and mammals per unit of their projection area exceeds the productive capacity of plants thousands of times (Gorshkov et al., 2000). Therefore, animals must consume plant-synthesized food on a territory hundreds and thousands of times larger than their projection area, eating in a mere hour what plants synthesize in a year. As a result, animals must constantly move around their feeding territory as an essential condition of their survival. At the same time, the quick consumption by animals of plant production accumulated by a local community inevitably leads to sharp fluctuations in these stocks because the subsequent regeneration of biomass proceeds at very different rates. These fluctuations, in their turn, overlap with fluctuations in inorganic matter—excreta eliminated by organisms after feeding into the environment—which also disrupts environmental stasis.

Therefore, the only way for large animals to exist within a very enclosed substance cycle is to limit their share of consumption of plant production. This consumption cannot exceed the natural fluctuations of organic synthesis, and the ingestion of biomass cannot leave a palpable trace in biotic communities. Correspondingly, this quota must be a negative function of animal size, in accordance with the observed use of primary production according to animal body sizes (Figure 14.1). Generally, this is achieved by the same means as for environmental biotic regulation, although as far as large animals are concerned, this is somewhat different. For example, when a given group of herbivores reaches excessive population density, there may be an increase in the number of plants endowed with burrs or characterized by repulsive taste. There are also plants with medicinal qualities or, conversely, containing toxins or narcotics, etc. Clearly, a community reacting in this manner to population fluctuations in large mammals will acquire an advantage—compared with others like them in a population. manifesting this capacity less markedly—and is more likely to survive in the competitive process of life. Therefore, it makes sense to assume that—in relation to biotic communities consisting of plants, fungi, and microorganisms—large animals constitute the same kind of component of the environment that they regulate as concentrations of biochemical substances in the soil, water, and air.

As for the predators, they are located at the top of the ecological pyramid and cannot go beyond numbers optimal for a given ecosystem, given the stability of their prey numbers. Therefore, under natural conditions predators cannot disrupt the ecological balance, and their role in the ecosystem consists of regulating prey numbers, as well as culling old and genetically defective individuals. It is probably no coincidence that the increase in genetic decay polymorphism in herbivores, which occurs under conditions of strong habitat disruption, is usually accompanied by the simultaneous rise in the number of predators. By the way, an analogous correlation is also observed between plants and plant-eating insects—when the former experience an increase in decay polymorphism, the population of the latter increases. Similar processes take place after forest fires, clear-cutting, and other strong environmental disruptions (Isaiev *et al.*, 1984; Morneau and Payette, 1989).

And so, the danger that large animals will destroy a given community is first of all based on their potential population growth beyond a certain critical threshold. Therefore, the normal behavior of higher animals with intact genetic programs is also directed at maintaining the stability of population density achieved by the limitation of births (in the case of food shortages as was demonstrated with the example of the tundra wolf in Chapter 2) and by the protection of their feeding territory on the part of the animals themselves (McNab, 1983). In undisrupted natural ecosystems this is achieved in particular by such mechanisms as (a) sound signals warning that a given territory is occupied, (b) animal migrations during periods of excessive population density, (c) the activation of parasites and predators that bring down population numbers, etc. Such intra-species and inter-species interactions are absolutely essential for the maintenance of competitive viability of biotic communities. After all, survival success is achieved not by species but rather by communities that ensure the effective control of life's mechanisms.

The question is why the biosphere needs large animals to begin with if it managed quite well without them for hundreds of millions of years, and even today the share of energy flow corresponding to these organisms is so small (less than 1%) that it cannot play a significant role in the general energetics of the biosphere. Nevertheless, the ubiquitous distribution of large animals (which is essential to practically every ecosystem) indicates that they somehow contribute to increasing ecological stability. One possible explanation for this phenomenon comes from V. G. Gorshkov who points out that disruptions of biota and the environment are quite random and irregular. Under conditions of a prolonged climax phase, the so-called "repair species" (see Chapter 10)—which are called upon to play a role during ecological disruption periods—end up gradually displaced from various communities. At such times they exist only as single individuals and practically cease to compete among themselves. This may lead to the disintegration of their genetic program that really comes into its own (in terms of its role in a community) only in a disrupted habitat.

In this manner, under the threat that the genetic information of the repair species will disappear or become distorted, biota must possess internal mechanisms for the regular disruption of ecosystems ensuring the maintenance of repair species populations at a certain minimally essential level. Apparently, this is the function assumed by the large animals that also belong for the most part to the repair species category and act as constant disruptors of ecosystems. Increases in the numbers of large animals in ecosystems subjected to forest fires, clear-cutting, and other destructive actions are an indirect indicator of their positive role when it comes to environmental "relaxation" and the return of the environment to an undisrupted state (Gorshkov, 1995; Gorshkov et al., 2004).

In light of the above we can see that, in addition to the ecological "mission" of large mobile animals, they also have a more modest role in the general energetics of the biota while the lead role is given to stationary organisms: plants, bacteria, and fungi. From this point of view, the biosphere can be viewed as a kind of energy machine meant for supplying large animals with food and oxygen, as well as for maintaining other vital parameters of their habitat. However, for this machine to function reliably, it is essential that its coefficient of efficiency does not exceed 1% while the remaining 99% of the general energy flow is used by species that produce the main contribution to the stabilization of the environment. Only when this condition is satisfied can biota ensure that the substance cycle is highly enclosed while the flow of synthesis and the flow of destruction are sufficiently well-matched—of the order of 10^{-4} (see Chapter 10).

And now let us return to the biosphere's carrying capacity—a kind of integral of the limit of human impact on the environment. As Holdgate (1994) once noted, numerous ecologists have broken their intellectual teeth trying to grapple with this concept since everything that has been said is directly related to carrying capacity. After all, humans also belong to the large animal category, and so the above limitation is fully applicable to us and to the animals in our care. And so, in order not to pull the carpet from underneath our own feet, we (and our entire economy) must fit into the framework of the same 1% energy corridor allotted in the biosphere to all large animals (Figure 14.2).

The 1% share of the general flow of energy in biota that we can use without undermining environmental stability is most easily expressed in net primary production values. Net primary production size can be expressed in units of mass of organic carbon (tons) or units of power (W_t) required for the amount of biomass produced by plants on corresponding territory per year, minus the expenditures for the maintenance and growth of the plants themselves. If the energy capacity of the Earth's entire biota is of the order of $100 \, TW_t$ ($10^{12} \, W_t$), then 1% of this amount will equal

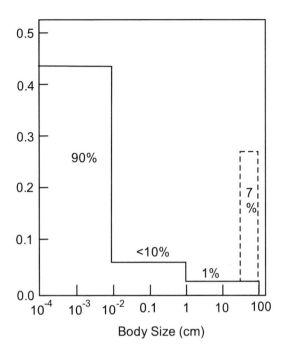

Figure 14.2. Share of consumption of net primary production allocated to heterotrophic individuals (bacteria, fungi, animals) of different body size in natural terrestrial ecosystems. The solid curve gives the universal distribution found in all unperturbed terrestrial ecosystems. The area enveloped by the solid curve is equal to unity. Numbers in percent give the relative input of different parts of the histogram. The dashed line describes violation of the natural distribution caused by present-day anthropogenic perturbation of the biosphere. The area under the anthropogenic peak (7%) corresponds to the food of humans, cattle fodder, and anthropogenic consumption of wood (from Gorshkov *et al.*, 2000, fig. 1.1).

approximately 1-2 TW_t. Evaluating the total mass of organic carbon, we arrive at a value of 1 Gt to 1.6 Gt C. Therefore, 1 TW_t to 2 TW_t (in units of power), or 1 Gt to 1.6 Gt C (in units of mass of organic carbon), provide us with a quantitative conception of the biosphere's carrying capacity: that is, its outermost limit that should not be exceeded by human civilization concerned with maintaining environmental stability.

However, today this objectively established human consumption limit for net primary production has already been exceeded by an order of magnitude. The 1 TW_t to 2 TW_t value corresponded to the capacity of civilization at the beginning of the previous century when we overstepped the forbidden threshold. It is no coincidence that atmospheric CO_2 began to grow quickly around 1900 as well (i.e., after human consumption of net primary production had increased beyond the admissible 1% barrier). By that time the Earth's population—which had reached 1.6 billion people—had already destroyed or seriously damaged natural ecosystems on 20% of the dry landmass. Therefore, 20% of the economically utilizable land (taking into account the period's technology) can also be viewed as the "geographic" equivalent of the biospheric disruption threshold. Furthermore, in line with conclusions stemming from the biotic regulation theory, global ecological equilibrium is seriously disrupted in the biosphere as a result of economic activity covering 25% to 30% of the dry landmass (Gorshkov, 1995).

But what is the significance of this long-exceeded 20% (or 1/5) of the dry landmass, given that the 20th century, in its mad industrial and demographic frenzy, went more than three times beyond this threshold, bringing the area of today's disrupted ecosystems to 63.8%? This too, is another indicator of the colossal transgression of the biospheric disruption threshold and the extent to which the compensatory capacity of the biosphere is close to exhaustion. This is also indicated by the disruption of the integrity of the biogen cycle (CO_2, nitrogen compounds, and phosphorus), and the progressive loss of biodiversity, as well as the shift of many (until recently) renewable natural resources into the category of non-renewable ones, not to mention many other phenomena discussed in Chapter 1.

However, if this first threshold—which is so critical for humanity—has already been transgressed, then a logical question to ask would be about the next threshold. This threshold would be far more serious, since at this stage, phenomena associated with biospheric deterioration would become irreversible, and the biosphere's capacity for self-regeneration would be lost for an indefinite period of time. At this point we would like to turn once again to the position of the so-called "technological optimists" who believe (almost religiously so?) in the limitless potential of scientific and technological progress. They feel that such advances, which have already resolved dangerous situations more than once, will successfully tackle the current ecological threat as well. Given the above considerations of biospheric and energy-based limitations, hopes for the creation of artificial environmental regulation mechanisms as a replacement for their natural counterparts appear, frankly, utopian.

Indeed, hardly anyone would contest that humanity is nowhere near to being able to regulate and control the environment with the same coefficient of efficiency and at the same energy level as that achieved by Earth's biota. In fact there is no guarantee that we will ever manage to master this global mechanism. The biota itself has arrived at this ability after a tremendously long evolutionary process. In order to gain a more concrete picture of our capacity for the creation of an artificial environment, let us recall the main aspects of the "expenses column" within the whole energy budget placed at the Earth's disposal by solar emissions that reach our planet.

The total power of solar emissions reaching the Earth beyond the atmosphere is approximately 10.5×10^6 kJ/m^2 per year of which about half or 5×10^6 kJ/m^2 per year reaches the surface of the planet directly. However, since most of this energy returns to the atmosphere through secondary radiation, only 1×10^6 kJ/m^2 per year reaches land vegetation and phytoplankton in mid-latitudes. However, 95% to 99% of this amount is immediately reflected, absorbed (which leads to heat generation), or used for the evaporation of water. Only 1% to 5% is taken up by green leaf chlorophyll and sets in motion the streaming of energy in Earth's biota (Green et al., 1984). Therefore, the entire energy capacity of biota ($100\,TW_t$) amounts to approximately 1/1,000 of the solar emissions that reach the Earth. And that is the figure constituting the basis for the stability of temperature, climate, and other environmental parameters.

Theoretically speaking, this is still not the limit, and biota could raise its energy capacity by another order of magnitude: for example, through the action of plants from the so-called C4 group that synthesize carbohydrates on the basis of the tetra-carboxylic acid cycle (Govindzhi, 1987). The latter include corn and sugar cane in particular. However, according to estimates, the present energy capacity of biota— which was apparently formed through evolution—has reached the biological stability limit of the modern climate. Beyond this, we will inevitably face unpredictable fluctuations of surface temperature (Gorshkov, 1990).

Therefore, even if we assume that humanity will one day find an unlimited source of "ecologically clean" energy (controlled thermonuclear fusion, powerful solar arrays set up in space, etc.) and gain control of environmental regulation, maintaining the same integrity of the substance cycle and achieving the same coefficient of efficiency as in the modern biosphere, we will still be unable to go beyond today's energy capacity of the biota without risking irreversible climatic disequilibrium. However, under these conditions we would have to use 99% of energy and labor expenditures of civilization for the maintenance of environmental stability (today, the cost of effective cleaning installations at times already reaches half the startup cost of a given enterprise). So what then would be left over for satisfying our own needs? It would be about the same as (or even less than) what we have at our disposal in natural biospheric conditions without expending a single calorie for the maintenance of environmental stability or giving a single thought to the ways in which this task is carried out by living biota. So much for the fabled ambition of getting by without nature.

References

Gorshkov, V. G. (1981). "Distribution of energy flow among organisms of different sizes," *Zhurnal obschei biologii* **42**, 417–429 [in Russian].

Gorshkov, V. G. (1990). *Biospheric Energetics and Environmental Stability [Itogi nauki i tekhniki, seria teoreticheskie I obschie voprosy geografii]*, Vol. 7. Moscow: VINITI [in Russian].

Gorshkov, V. G. (1995). *Physical and Biological Bases of Life Stability: Man, Biota, Environment*. Berlin: Springer-Verlag.

Gorshkov, V. G., V. V. Gorshkov; A. M. Makarieva (2000). *Biotic Regulation of the Environment: Key Issue of Global Change*. Chichester, U.K.: Springer/Praxis.

Gorshkov, V. G.; A. M. Makarieva; V. V. Gorshkov (2004). "Revising the fundamentals of ecological knowledge: The biota–environment interaction," *Ecological Complexity* **1**, 17–36.

Govindzhai, D. H. (Ed.) (1987). *Photosynthesis*, Vols. 1–2. Moscow: Mir.

Green, N. P. O.; G. W. Stout; D. J. Taylor (1984). *Biological Science*, Vol. 2 (edited by R. Soper). Cambridge, U.K.: Cambridge University Press.

Holdgate, M. W. (1994). "Ecology, development and global policy," *Journal of Applied Ecology* **31**(1), 201–211.

Houghton, J. T.; L. G. Meira Filho; B. A. Callander; N. Harris; A. Kattenberg; K. Maskels (Eds.) (1996). *Climate Change 1995: The Science of Climate Change*. Cambridge, U.K.: Cambridge University Press.

Isaiev, A. S.; R. G. Khlebopros; L. V. Nedorezov; Yu. P. Kondrakov; V. V. Kiseliov (1984). *The Dynamics of Insect Population Numbers*. Novosibirsk: Nauka SO [in Russian].

McNab, B. K. (1983). "Energetics, body size, and the limits to endothermy," *Journal of Zoology* **199**(1), 1–29

Morneau, C.; S. Payette (1989). "Postfire lichen–spruce woodland recovery at the limit of the boreal forest in northern Québec," *Canadian Journal of Botany* **67**, 2770–2782.

Ricklefs, R. E.; G. L. Miller (2000). *Ecology*. Columbus, OH: W. H. Freeman.

Sakharov, A. D. (1990). *Peace, Progress and Human Rights: Articles and Speeches.* Leningrad: Svetsky pisatel' [in Russian].

Zavarzin, G. A. (2000). "The non-Darwinist area of evolution," *Vestnik RAN* **70**(5), 403–411 [in Russian].

15

The starting conditions of sustainable development and the preservation of ecosystems by country and continent

The key component of the ecological crisis—View from space: the state of natural ecosystems from the Arctic to the Antarctic—World centers of environmental stabilization—How much anthropogenic carbon can be absorbed by ecosystems on land and in the oceans—Possible timeout for nature—The A-, B-, and C-students: countries assessed in terms of ecological system preservation—Global centers of environmental destabilization—Southeast Asia and the Atlantic countries: two poles of ecological ills—Six million square kilometers of land to be returned to nature—What can the developed countries do to curb population growth in the Third World—Sustainable development and world partnership: ecological donor and recipient countries—The developed world and the ecological spaces of Russia

A researcher trying to postulate possible scenarios and prospects for future development runs up against the wall of historical inscrutability. Ecologists would feel much less constrained if science could demonstrate definitively and convincingly to the overwhelming majority what inverse relationships should come into play as the ecological crisis deepens and what alarming symptoms coming from modern civilization ought to be interpreted as clear harbingers of disaster.

In fact, almost everyone is talking about the local and regional consequences of environmental degradation and pollution leading to desertification or rising rates of disease. As for the global manifestations of this process, they are still largely rather vague. In part this has to do with the impossibility of determining with certainty the numbers of people who have died or have become handicapped as a result of the global nature of the current situation. Furthermore, we still lack clear data on the way that the critical overpopulation of the planet and its deteriorating ecology influence social behavior and modern institutions. There are some indications that the present conflict in Sudan may be a consequence of desertification (i.e., a struggle for dwindling water resources). Similarly, in his recent book tellingly entitled *Collapse: How Societies Choose to Fail or Succeed*, Jared Diamond argues that the

Rwandan genocide of 1994 was attributable to overpopulation, with ethnic tensions acting merely as catalysts for a problem caused by the insufficiency of land resources to feed the farmers of the tiny country. Quoting Gérard Prunier, Diamond sums up this position as follows: "The decision to kill was of course made by politicians, for political reasons. But at least part of the reason why it was carried out so thoroughly by the ordinary rank-and-file peasants in the *ingo* [= family compound] was feeling that there were too many people on too little land, and that with a reduction in their numbers, there would be more for the survivors." Perhaps the most striking feature of these murders is that they often occurred *within* families (Diamond, 2005, pp. 324–326).

However, such piecemeal evidence or conjecture—in the case of the Rwandan example, as Diamond admits himself, the overpopulation–genocide link has been challenged by various researchers—has not yet coalesced into a global picture that pulls all the loose ends together. Even more so, when it comes to forecasting future trends, many scientists tend to shy away from definitive statements. In this connection we may be dealing with a Catch-22 situation: namely, if we do end up recognizing unmistakable signs of impending disaster, will humanity be in a position to reverse the process it has unleashed in the biosphere? However, even to reach this recognition stage, we have to realize as a planetary community that the main problem is not environmental pollution or anthropogenic global warming—as serious as these threats may be. We have been trying to argue in this book that the most harmful consequence of human economic and technological activity (especially in the past century or so) is the destruction of natural ecosystems over huge areas of land and in the water areas of semi-enclosed seas and coastal oceanic zones. It is the drastic weakening of the environment-shaping and stabilizing function of biota that constitutes the greatest threat to the biosphere. Therefore, if there is any hope to prevent the collapse that Diamond depicts in his book, it lies in our ability to rely on the natural potential of living biota.

This is the idea behind sustainable development within the theory of environmental biotic regulation. If the real goal in question is the weakening of anthropogenic pressure to a level corresponding to the biosphere's carrying capacity, the key issue is not just stopping our advance against nature, but rather pulling back and actually *freeing up* some of the territory we have occupied in order to let the biosphere recover (cf. Chapter 7). Only then can biota resume its mission of stabilizing the planetary environment. The monumental nature of the challenge before us is not just our obsession with progress and science, as well as the short-term outlook of the political process. The other obstacle is the extreme unevenness of the starting conditions characterizing various countries and regions of the world as they attempt to bring about sustainable development. It is sufficient to compare certain African or Asian states which still cling to aspects of late feudalism with the information-based societies of Western Europe and North America in order to grasp the depth of the chasm that needs to be overcome by the international community on its way to the solution of key global problems.

There is, however, a criterion allowing us to juxtapose and compare different countries regardless of their cultural differences, financial flow, development of industrial infrastructure, and resource wealth. This is the degree of ecological preservation—an asset far more valuable than mounds of diamonds or gold reserves in bank vaults. When viewed from this angle, the most "advanced" regions of the world are those where wild nature has not yet been destroyed. They are the keepers of the planet's real wealth—whether or not they possess a large per capita share of microprocessors and cellphones. And so, it is from this position that we will now attempt to evaluate the starting conditions of various countries on the road to sustainability.

In order to acquire a sense of the distribution of ecosystem disruption according to countries and continents, we need to consider satellite data that paint a vivid cartographic picture of the Earth's biosphere. Today territories with intact ecosystems form 51.9% of the dry landmass or 77 million km^2. However, much of that area—20 million km^2—corresponds to glaciated or exposed rocky surfaces in such places as the Antarctic, Greenland, the Himalayas, etc. (Hannah *et al.*, 1994). Although glaciers, just like the hydrosphere in general, play an important role in stabilizing the Earth's climate, their biological productivity is close to zero as is consequently their potential for biotic regulation of the environment. Therefore, only 57 million km^2 or 37% of the planet's dry landmass is biologically productive, and the distribution of intact ecosystems is highly uneven.

Alongside relatively small islets of undisturbed wild nature ranging in area from 0.1 million km^2 to 1 million km^2, we find a few massive territories covering many millions of square kilometers. The latter are the so-called centers of environmental stabilization (see Figure 15.1), two of which are located in the northern hemisphere and one in the southern hemisphere:

- The North Eurasian Center (11 million km^2) including Northern Scandinavia and the north European part of Russia, as well as most of Siberia and the Far East (excluding their southern regions).
- The North American Center (9 million km^2) including the northern part of Canada and Alaska.
- The South American Center (10 million km^2) including Amazonia along with the adjacent mountainous territories.

It is evident that most of these vast woodlands and wetlands are concentrated mainly in three countries: Russia (3.45 million km^2), Canada (3.43 million km^2), and Brazil (2.28 million km^2). This is 68% of the planet's total reserves. Altogether, virgin forests and wetlands cover around 13.5 million km^2 of land, which (according to different estimates) corresponds to 40% to 44% of today's populated areas (Bryant *et al.*, 1997). Like the whales of antiquity which bore the weight of the Earth on their backs, these preserved ecosystems, as well as the World Ocean with its (so far) relatively undisturbed ecosystems, make a decisive contribution to maintaining the stability of the biosphere, allowing it to withstand more or less successfully the growing anthropogenic pressure.

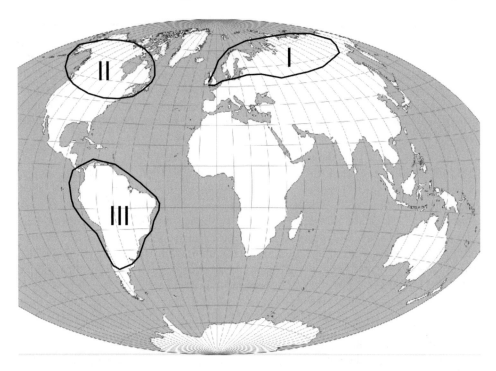

Figure 15.1. Centers of global environmental stabilization: I—North Eurasian, II—North American, III—South American.

It is common to view the forests as the lungs of the Earth; however, it would be far more appropriate to compare them to the planet's kidneys since they act as drainage for the elimination of excess atmospheric biogens (including CO_2) from circulation. The humus layer and swampy bogs are referred to by some authors as "eternal" carbon traps where, as in the case of marine deposits, the carbon can remain undisturbed indefinitely under appropriate conditions (Vompersky, 1994). However, this applies only to undisrupted ecosystems which include, for example, frontier or climax forests with their small accretion of wood and large numbers of fallen, rotting trunks. At the same time such forests possess a high level of biological productivity ensuring nutrients and reproductive capacity for all kinds of natural communities that are responsible for the maintenance of environmental stability. Disrupted ecosystems (e.g., forests that are periodically logged partially or completely) act very differently. As research data demonstrate, reclaimed lands are characterized by biota that does not absorb excess atmospheric CO_2. Instead the opposite is the case: that is, biota acts as a source of additional CO_2 emissions while the reserves of carbon accumulated in the forests diminish—at a rate of $1.1\,\mathrm{Gt\,C\,yr^{-1}}$ according to FAO data (*Global Forest Resources Assessment*, 2005). Furthermore, as was noted earlier (Chapter 10), biota has developed a huge capacity for the production and destruction or organic matter, which has the hidden dangerous potential for

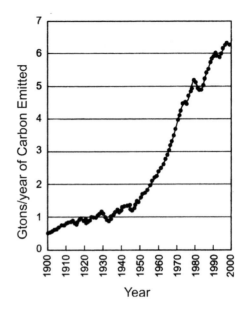

Figure 15.2. Growth of CO_2 emissions worldwide as a result of fossil fuel burning (Gt/yr) (Worldwatch Database, 2000).

rapid environmental degradation through the disruption of the integrity of the biological substance cycle.

Measurements taken since 1958 at various observation sites around the world indicate that the concentration of atmospheric CO_2 is constantly growing (see Figure 15.2). Analysis of air bubble composition from Antarctic ice cores (Friedli *et al.*, 1986; Staffelbach *et al.*, 1991; Raynaud *et al.*, 1993, Petit *et al.*, 1999) makes it possible to picture the dynamics of atmospheric CO_2 concentration before and after the beginning of global atmospheric disruption that coincided with the beginning of the Industrial Revolution at the end of the 18th century. For example, ice core research has shown that pre-industrial CO_2 concentration was approximately 280 ppmv (parts per million by volume), and this level had remained virtually constant for several preceding millennia (Siegenthaler and Oeschger, 1987; Lorius and Oeschger, 1994). However, by now the atmospheric CO_2 concentration has already reached 370 ppmv (Siegenthaler and Sarmiento, 1993; Etheridge *et al.*, 1996): that is, it has surpassed the pre-industrial level by 28%. And this growth began even before the large-scale use of organic fossil fuels that overlapped with carbon emissions provoked by land utilization. However, from that point on through the end of the 19th century, the maintenance of biospheric stability was ensured mainly by the ecosystems of the World Ocean whose compensatory potential reached its critical limit by the late 19th century. After that, the process of global environmental change began.

To date, several attempts have been made to calculate the balance of CO_2 (in the form of carbon) in the biosphere based on the law of conservation of matter and stoichiometric (volume–mass) O_2–CO_2 ratios for the main pools of sources and discharge channels of anthropogenic carbon. A widely used analysis was undertaken by Houghton *et al.* (1996). The common position is that changes in carbon content

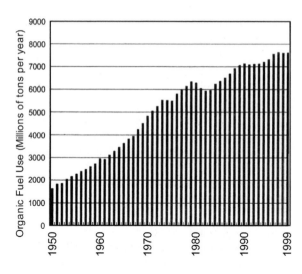

Figure 15.3. Worldwide use of mineral sources of organic fuel (millions of tons of petroleum equivalent) (Worldwatch Database, 2000).

take place in four interconnected media: the atmosphere, the oceans, land, and (in the past 200 years) hydrocarbon fossil fuel deposits (see Figure 15.3). The combination of these forms of flux should add up to zero: that is, the sources of CO_2 should be compensated by its outflow. Apart from hydrocarbon fuel burned in power plant furnaces and engines, large amounts of CO_2 reach the atmosphere in the process of cement production, the burning of associated gas, as well as from agricultural activity accompanied by the destruction of biomass (logging, humus destruction during tillage, etc.) The atmosphere, the World Ocean, and undisturbed ecosystems on dry land, act as repositories for accumulating CO_2.

The cumulative sum of CO_2 emissions from fossil fuel combustion since the beginning of the Industrial Era is estimated to be $7\,Gt\,C\,yr^{-1}$. The current rate of CO_2 accumulation in the atmosphere is of the order of 2.2 Gt/yr. Finally, based on the ratio of $^{13}C/^{12}C$ isotopes in oceanic water and in the air, it has been possible to determine the rate of CO_2 absorption by the ocean through physicochemical processes—the diffusion of excess CO_2 through the surface of the air–water discontinuity—and the leveling of its concentration in accordance with Henry's Law. This rate is estimated to be $2.6\,Gt\,C\,yr^{-1}$ (Losev, 2001). It is somewhat more complex to attempt an estimate of: (a) carbon emission from land utilization which is determined from biomass reduction on a given territory, and (b) CO_2 absorption by undisrupted ecosystems in the oceans and on land.

In this connection, the absence of precise methods for tracking the production and destruction of organic matter creates certain difficulties that may introduce errors. For example, the influx of carbon from land-based biota can take place through the destruction of natural (primarily forest) ecosystems and that of the soil layer. However, while forests—after their exploitation ceases (provided it has not gone too far)—retain their capacity for the regeneration of plant biomass and begin to absorb the accumulated excess atmospheric CO_2 (Gorshkov, 1980; Vitousek *et al.*, 1986; Wofsy *et al.*, 1993), this cannot be said about the soil—the main reservoir of biogens on dry land. This is because accumulation of carbon in the soil takes place extremely slowly while its loss can take place very rapidly. Thus, the average loss of soil carbon during the transition to soil cultivation amounts to around 30%, and in the Tropics it can even reach 70%. In other words, the regeneration of disrupted ecosystems is normally accompanied by the reduction of carbon content in the soil, with resultant incomplete compensation.

The following is a more precise balance of the global carbon cycle calculated by Gorshkov *et al.* (2000) and taking into account the action of the oceanic biotic pump and the contribution made by disrupted and undisrupted ecosystems on dry land (Figure 15.4). The basis of this balance includes data from the yearly flow of net primary production in the biosphere as a whole (of the order of $1,000 \, \text{Gt C yr}^{-1}$), as well as in the World Ocean ($40 \, \text{Gt C yr}^{-1}$) and on land ($60 \, \text{Gt C yr}^{-1}$) (Falkowsky and Woodhead, 1992; Lurin *et al.*, 1994). The latter, in its turn, is subdivided into two unequal parts corresponding to disrupted and undisrupted biota on dry land, since the two act differently when it comes to the absorption of atmospheric CO_2. Thus, 24% of the total biospheric flow of net primary production formed by biota on land corresponds to its share of undisrupted ecosystems whereas 36% corresponds to economically reclaimed territories. The former, in accordance with *Le Chatelier's Principle*, fix the carbon accumulated in the atmosphere by means of organic synthesis, while the latter are sources of emission.

Another global source of CO_2 emissions—to which most authors tend to attribute a key role—is fossil fuels. In this connection we ought to note that the scientific community has not managed to agree on what happens to so-called "lost" carbon formed after the following are deducted from total carbon emissions:

- CO_2 emitted through the burning of fossil fuels ($5.9 \, \text{Gt C yr}^{-1}$);
- CO_2 accumulated in the atmosphere ($2.2 \, \text{Gt C yr}^{-1}$);
- CO_2 absorbed by the World Ocean through physicochemical processes ($2.6 \, \text{Gt C yr}^{-1}$).

In this connection researchers often underestimate the role of oceanic biota, as well as the absorption and emission of carbon by disrupted and undisrupted ecosystems on land. We would consider this to be a serious mistake; first of all because in remote geological epochs, when there was still no land-based life, it was specifically by means of the oceanic biotic pump that huge amounts of carbon were taken out of the atmosphere and buried in bottom sediments as attested to by paleological data. The second reason is that the undisrupted ecosystems on land—with their humus, tundra swamps, or the bogs of boreal forests which act as long-term (some authors even use the word "eternal") carbon reservoirs—are in fact analogous to oceanic systems with their bottom sediments and poorly mixed cold deepwater.

Alas, this cannot be said about artificial agrocenosis: intensively exploited forests, arable land, pastures, etc. Here the disruption of the biological cycle—the discrepancy between organic synthesis and organic destruction—can reach tens of percent. The emissions of carbon into the atmosphere from disrupted ecosystems on land, as was already pointed out in Chapter 10, are of the same order of magnitude as

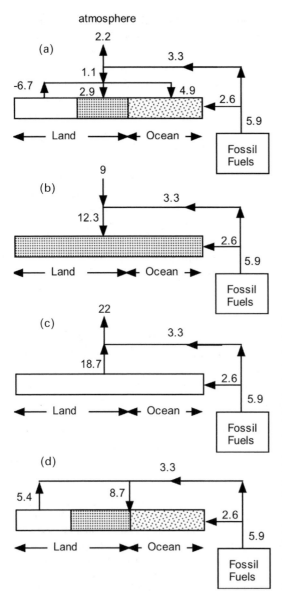

Figure 15.4. Global change and natural biota. Global carbon cycle change: (a) for the modern state of the biosphere, (b) for a completely unperturbed biosphere, (c) for a completely perturbed biosphere, (d) for stopping global change. Shaded boxes indicate natural, unperturbed biota. Blank boxes indicate anthropogenically perturbed biota. Numerical figures near arrows are carbon fluxes in gigatons per year. Numerical figures in boxes are values of net primary production in gigatons per year of carbon. The carbon flux of 2.6 Gt/yr represents absorption of atmospheric carbon by the oceans (based on Gorshkov *et al.*, 2000, fig. 6.6).

the intensity of carbon emissions from the burning of coal, petroleum, and gas (Houghton, 1989; Watts, 1982; Rotty, 1983), according to Gorshkov's calculations, are even greater than emissions from burning (see Figure 15.4).

For now, the capacity of oceanic biota ($4.9\,\mathrm{Gt\,C\,yr^{-1}}$) and that of ecosystems on land ($2.9\,\mathrm{Gt\,C\,yr^{-1}}$)—see shaded areas in Figure 15.4—is sufficient for the absorption of CO_2 emissions coming into the atmosphere from territories reclaimed by civilization. As for fossil fuels, the same diagram shows that $2.6\,\mathrm{Gt\,C\,yr^{-1}}$ of CO_2 formed in this manner ends up in the oceans as a result of physicochemical absorption and another $1.1\,\mathrm{Gt}\ C\,yr^{-1}$ is absorbed by oceanic biota and intact ecosystems on land while the remaining $2.2\,\mathrm{Gt}\ C\,yr^{-1}$ is accumulated in the atmosphere.

At the same time, Figure 15.4 depicts two scenarios: the balance of the global carbon cycle under the conditions of complete natural ecosystem recovery (Figure 15.4b: ocean and land shaded completely) and in the case of complete biospheric reclamation by man (Figure 15.4c: ocean and land unshaded). In the first scenario the undisrupted biosphere could absorb all of the carbon emitted into the atmosphere even with the current volumes of fossil fuel use. In this case atmospheric CO_2 concentration would begin to diminish at the rate of $9.0\,\mathrm{Gt}\ C\,yr^{-1}$, which would make it possible to bring the biosphere back to its pre-industrial state within a few decades. On the other hand, under the conditions of complete biospheric reclamation, carbon emissions would equal $18.7\,\mathrm{Gt}\ C\,yr^{-1}$, and the concentration of atmospheric CO_2 would reach catastrophic proportions—of the order of $22.0\,\mathrm{Gt}\ C\,yr^{-1}$. In such a situation, even the complete renunciation of fossil fuel use would still fail to save the day since this would merely reduce the overall CO_2 emissions to $16.0\,\mathrm{Gt}\ C\,yr^{-1}$.

Finally, in order to halt the accumulation of atmospheric CO_2 given today's levels of fossil fuel burning (i.e., "freeze" the present state of the biosphere for a while), humanity would need to lower net primary production from 36% to 29%. Put another way, we would have to vacate approximately 7% of reclaimed territories (see Figure 15.4d). In this case the CO_2 emissions from reclaimed land would be $5.4\,\mathrm{Gt}$ $C\,yr^{-1}$ while undisrupted oceanic biota, as well as intact and resuscitated land-based ecosystems, would be able to supply the outflow of $5.4\,\mathrm{Gt}$ of inorganic carbon, including the $3.3\,\mathrm{Gt}\ C\,yr^{-1}$ which, once we deduct the carbon absorbed by oceanic waters, are emitted on a yearly basis into the atmosphere through the burning of fossil fuels. Essentially, this 7% of vacated land would amount to the decision not to use 40% of forests that are currently part of our economic activity. All the corresponding wood would have to be replaced by artificial material. Such would be the price to pay for the timeout that humanity would give nature in exchange for the time required to solve our demographic challenge, as well as the energy problem and other ecologically significant issues.[1]

---

[1] At this point we should note the fundamental difference between the regeneration of natural ecosystems and the creation of artificial plantings. The former are not random communities of organisms. Their unity evolved in order to maintain environmental stability: that is, these communities, in line with Le Chatelier's Principle, can compensate for periodic environmental disruptions. As for artificially maintained biological systems, they normally consist of randomly selected (from nature's point of view) species. As such, they are normally not only unable to ensure their own stability, but also, given their high biological productivity, themselves act as a constant source of global environmental disruption. In this light, the complete absence of vegetation on a given territory is actually more preferable than an artificially created and maintained ecosystem. And so, we are not talking about artificial resuscitation, but rather natural communities capable of independent (natural) regeneration on territories vacated by human beings. This does not imply that people would have no access to such lands; however, those who would like to come close to nature in these areas would have to use nothing but their muscles to move around, build, or do any other work.

Whether or not these goals are realistic will be dealt with below; for now, however, let us note that any contribution to their realization must not only be commensurate with the economic capacity of every country, but should also reflect the current degree of natural system preservation on a given territory. In other words, this is what one could call ecological fairness. While the World Ocean can be viewed as a *common* good, ecosystems on dry land (which also act as runoff areas for anthropogenic carbon) fall under the jurisdiction of specific countries and peoples that have managed to preserve them in one way or another. In this connection, all the world's countries can be grouped into three categories according to their social and natural parameters:[2]

(1) Good starting conditions for the transition to sustainable development (area of undisrupted ecosystems exceeds 60% of territory): 8 countries or 5.5% of the world's states.
(2) Poor starting ecological conditions (area of undisrupted ecosystems below 10% of territory): 91 countries or 62% of the world's states.
(3) Mid-level starting ecological conditions (area of undisrupted ecosystems equals 10% to 59% of territory): 47 countries or 32.5% of the world's states (Danilov-Danil'yan and Losev, 2000).

No one can compete in this connection with the world's two biggest countries (territorially speaking): Russia and Canada. They occupy 35% of the global landmass with undisrupted ecosystems. If we consider territories with the most ecologically productive forest ecosystems, the virgin forest area of Russia alone amounts to more than a quarter of the world's resources. We can now see the exclusive role played by these two countries in the preservation of the planet's biosphere. Six other countries are part of the same relatively well-off group: Algeria, Mauritania, Botswana, Lesotho, Guyana, and Surinam. However, their roles in global ecodynamics is incomparably more modest, both in light of their smaller territories and because the undisrupted ecosystems of the largest two countries of this group—Algeria and Mauritania—consist mainly of deserts and semi-deserts.

As for the countries with low-level starting ecological conditions (area of undisrupted ecosystems below 10% of territory), most of them are concentrated in three global centers of environmental destabilization (Figure 15.5):

(1) Europe, including central, western, and eastern areas (but not Norway and Iceland), as well as the European part of Russia, Ukraine, Belarus, and the Baltic States—with a total area of 8 million km^2 and 8% of natural ecosystems intact.
(2) North America, including the United States (but not Alaska), southern and central Canada, and the north of Mexico—with a total area of 9 million km^2 and less than 10% of natural ecosystems intact.
(3) Southeast Asia, including the Indian Subcontinent, Malaysia, Burma, Indonesia

---

[2] The following does not include a number of countries for which there are no available or reliable statistical data.

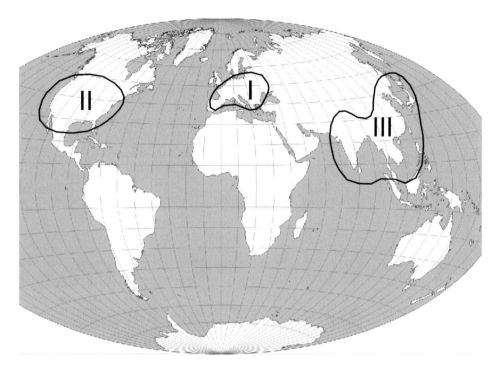

Figure 15.5. Centers of global environmental destabilization: I—European, II—North American, III—Southeast Asian.

(but not the island of Sumatra), China (but not Tibet), Japan, the Korean Peninsula, and the Philippines—with a total area of 7 million km^2 and less than 5% of natural ecosystems intact.

It is easy to see that one pole of the three centers in question consists of Europe's and North America's industrialized states while the other one is made up of developing countries (excluding Japan) with high population growth rates and largely low standards of living. Both the European and the Asian centers are the most ancient areas of civilization expansion, which implies that powerful anthropogenic pressures have been acting on the natural environment for many centuries. For example, the forests of the Apennine Peninsula were wiped out during the period of the Roman Empire while the same thing occurred in Western and Central Europe in the Middle Ages through the intensive development of agriculture, urban expansion, and the production of charcoal for iron smelting.

The discovery of America and the subsequent Industrial Revolution intensified the process of natural ecosystem destruction in the Old and New Worlds, and Great Britain can be viewed as a textbook example of this phenomenon. Simply put, the forests of England were "eaten" by sheep. As early as the 18th century, the textile industry required more and more wool, so sheep pastures were expanded at the expense of forests. Furthermore, timber was used for the construction of the English fleet. The result was that Great Britain ended up almost entirely deforested with only a few remnants of forested areas covering only 12% of its territory—mainly in the highlands of Scotland. The famous English oak forests are now to be seen only as part of sets for movies about Robin Hood whose "men in green" would find very little camouflage in today's landscape.

The unfavorable ecological situation in industrial countries usually goes hand-in-hand with high population density. Thus, in Western Europe, excluding Iceland and the countries of the Scandinavian Peninsula, population density is between 57 (Ireland) to 395 (The Netherlands) people per km^2. In Japan the figure is 337 while in the United States it is 30 people per km^2 (CIA, 2004). Therefore, freeing up geographic space for ecological needs constitutes a major problem for most of these areas.

Another important characteristic of developed countries, which supply around 80% of the world's industrial production exports, is the use of other countries' ecological spaces—primarily those of the Third World. The latter is the source of large raw material influx and the host of resource-intensive sectors, as well as the chemical industry, transferred from the developed world. Given all this, the industrialized world creates two-thirds of the world's waste on its territory and is the main emitter of anthropogenic CO_2 although this camp has now been joined by some of the biggest developing countries, such as China and India (see Figure 15.6).

Still worse are the starting conditions of the countries at the opposite pole (the Southeast Asian center of environmental destabilization) which, according to the latest data, is the biggest climate destabilizer on the planet—primarily as a result of CO_2 emissions through land use (Zalikhanov et al., 2006). These are primarily developing countries that have made a substantial leap forward toward industrialization and the use of new technologies. Still, they are characterized by largely low (less

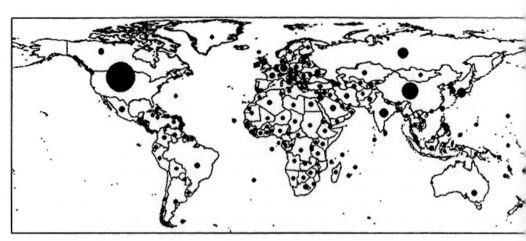

Figure 15.6. Emissions of CO_2 from the burning of fossil fuels by individual countries in 2000; relative size of black circles correspond to CO_2 emission volumes for a given country's territory.

than $600) and intermediate ($600 to $1,000) levels of per capita GDP. Here too, population density is very high—from 72 (Malaysia) to 982 (Bangladesh) people per km^2—and there are virtually no undisrupted ecosystems left. Only small parts of the original tropical rainforests are still standing in Malaysia, Indonesia, Thailand, and Pakistan where they cover 9%, 8%, 7%, and 5% of the respective territories. The tropical rainforests of Southern China have been almost entirely wiped out within the span of human history. Today, only the first few percent of their initial territories have been preserved (*World Resources, 1990–1991*, 1990).

However, perhaps the most essential feature of most countries in this region is poverty. Three of the most populated countries in the world—India, Pakistan, and Bangladesh, where altogether about 2.3 billion people live—are also among the states with the lowest GDP (less than $500 per capita). Put together with China ($989 per capita), these countries make up more than 40% of the Earth's population, so that poverty and overpopulation are associated with profound ecological ills. When the lives of huge human population masses waver on the edge of survival, we observe the collapse of the ethical constraints holding back the assault on nature. This is particularly vividly illustrated by the example of Bangladesh where there are no longer any remaining territories with intact ecosystems, and India is not much better off—with only 1% still intact. To quote Indira Gandhi, "poverty is the worst form of pollution" (UNDP India, n.d.).

In regard to the social unrest plaguing such states as Pakistan or Bangladesh, Jared Diamond points out that these sociopolitical failures "prove to be measures of environmental and population pressure, such as high infant mortality, rapid population growth, a high percentage of the population in their late teens and 20s, and hordes of unemployed young men without job prospects and ripe for recruitment into militias" (Diamond, 2005, p. 516). A seemingly endless series of caste, international, and religious conflicts as well as even genocide (as in the case of Bangladesh in the 1970s) are all indicative of mounting dangerous social energy which is likely to remain a source of instability for a long time to come in this overpopulated part of the planet. Only China appears to have found its own "unique" solution to this extremely complex problem by first "letting out the steam" through the Cultural Revolution in the 1960s and then resorting to a large-scale clampdown which was followed by widespread famine.

Countries with intermediate ecological starting conditions (area of undisrupted ecosystems in the range 10% to 59% of territory) correspond to around 60% of the world's potential of untouched wild nature. There are 47 such states—mainly from among the developing (and in some cases extremely economically backward) countries—most of which are located in the tropical and subtropical latitudes. Asia has 11 of them, South America 7 and Africa (mainly western and central) 21. Their role is particularly important when it comes to the preservation of the planet's tropical belt—our most valuable and irreplaceable eco-resource. Unfortunately, it is the tropical rainforests that are currently experiencing the greatest amount of anthropogenic pressure: that is, merciless logging by international companies that sometimes rely on corrupt local regimes and the desperately poor populations for logging leases: "Under these circumstances, rape-and-run will continue to be good business until the companies start to run out of unlogged countries" (Diamond, 2005, p. 472). Deforestation is progressing the fastest in Argentina (1,550,000 ha/yr) and in Brazil (2,323,000 ha/yr). Meanwhile, the ratio of forest regeneration areas to forest destruction areas is 1:100 and sometimes even 1:1,000 (Brazil, Indonesia) (*World Resources, 1990–1991*, 1990). Some former exporters of tropical forest materials (e.g., Malaysia) are now even forced to import wood for their own wood-processing needs. And so, while one can perhaps still view the need to preserve the islets of wild nature—the stem cells of future ecosystems—as the internal national goal of various ecologically ill countries, the problem of dwindling tropical forests is an acute collective problem of modern humanity as a whole.

As "cruel" as the foregoing may sound, knowledge about the current situation is the nucleus of any hope for a solution—however difficult that solution may be. It is only ignorance, complacency, or self-delusion that guarantee the realization of the direst doomsday predictions bombarding us from so many sources. As has been stated, just to "freeze" the biosphere in its current disrupted state—given unchanged volumes of fossil fuel burning and existing population size—humanity has to free up no less than 7% of today's reclaimed territories. If the total area occupied by modern civilization is about 86 million km^2, then we require the resurrection of natural ecosystems on approximately 6 million km^2 of land. What does the 6 million km^2 figure represent? Let us recall, for example, that in Russia all the forests under special protection cover 0.14 million km^2, its undisrupted forests and wetlands occupy 2.89 million km^2, and its entire forested domains stretch over 11.18 million km^2 (Aksenov *et al.*, 2002). Finally, the total area of the world's forest plantations is 1.79 million km^2 (*Global Forest Resources Assessment*, 2005).

So, is there a point in even discussing such an unimaginable number? Is it as far-fetched as planning the resettlement of humanity on the Moon? The daunting scale of the problem corresponds to the amount of harm done by man to nature in the second half of the 20th century alone. And yet, however daunting it may be, doing nothing is tantamount to collective suicide. Now, as is well known in the exact sciences, a challenge that at first seems insurmountable can turn out to be surmountable if the problem is divided into subproblems and stages. Therefore, let us begin by approaching the problem as the next stage in sustainable development whose most urgent goals should be

(1) Preservation of territories with natural systems still intact.
(2) Stabilization of the Earth's population.

Both issues—however unprecedented the degree of their complexity may be—are surely the most pressing ones, especially for poor developing countries with backward economies and nature pushed to the limit. With respect to the question of population, Dolnik (1992) writes: "Simply because every person can ... consciously control their fertility, one should not conclude that things are as simple at the level of populations ... The fertility of populations is controlled by populational mechanisms which operate quite apart from (and often despite) our collective consciousness." Using the example of countries with falling birthrates where attempts were made to boost or maintain natality by means of state policies (e.g., France or the former Soviet Union), Dolnik demonstrates how the same incentives temporarily raised the birthrates among the peoples of Central Asia (where they were already high to begin with) but failed to influence the stable population numbers in Russia or halt the decline in the population of Estonia, Lithuania, and Latvia. The author concludes: "What we need is not forced measures, but rather the creation of conditions in society whereby every person is maximally free from the influence of others when deciding how much progeny to have; possesses the relevant information from childhood; and has access to modern means for hindering, as well as facilitating conception" (Dolnik, 1992).

In other words, voluntary family planning is the only acceptable method and certainly the optimal method for stabilizing the population of countries undergoing the stage of demographic explosion—all the more so because this approach is not likely to create any social upheaval. It is probably not for nothing that the excessively harsh birthrate reduction policy implemented by the government of Indira Gandhi in 1977 led to serious discontent, unrest, and the prime minister's resignation. Admittedly, the same policies have been more successful in countries with different political systems and traditions (e.g., Indonesia or China). However, in the case of the latter, there are indications of a sexual imbalance stemming from the enforced one-child policy, which may have unforeseen negative consequences of its own (Gittings, 2002).

At the same time, however welcome voluntary family planning may be, it cannot work as long as the future of children in poor countries remains uncertain and fraught with danger in terms of health and personal safety. In the face of high infant and child mortality, the birthrate will continue to remain high while people have "supplementary" children as a kind of insurance policy meant to offset possible losses of progeny: that is, poverty and sociopolitical instability have a built-in mechanism of demographic excess. Therefore, while well-delivered sexual education programs and the equality of women with respect to questions of procreation (an issue most urgent for Moslem countries) should become priorities in the policies of administrations governing countries with rising population numbers, the Third World requires assistance on numerous fronts—political, social, and economic—before voluntary family planning becomes a reality.

Equally important is the need for cooperation between the haves and the have-nots when it comes to the fate of intact ecosystems—in the first place, the practically unrecoverable tropical rainforests. As pointed out earlier, the logging of these forests takes place not only for the purpose of wood marketing, but also to clear agricultural land, as well as for the satisfaction of basic daily needs: heating, cooking, etc. Given the rising population numbers in the affected areas, this process should not be underestimated. Therefore, increased foreign aid aimed at installing and servicing independent energy supply equipment or so-called mini-power plants (with a wattage of up to 100 kW) in remote impoverished areas could go a long way toward resolving the conflict between daily survival and environmental protection. This is a question of priorities in foreign aid policy. Money can be spent on different projects and in different amounts, but in light of the foregoing, the issue of land clearing and the energy supply should be at the top of the developed world's "good deeds" list—all the more so because there are no merely local problems when it comes to the planet's forests.

Remaining on the subject of international partnerships and cooperation, let us reconsider the 6 million km^2 of natural ecosystems that must be restituted to nature as a minimal condition for stabilizing the present situation. It is impossible to divide this territory fairly among all the 146 countries mentioned above. After all, this group includes Holland and Bangladesh—with their almost entirely consumed ecological spaces and extremely high population density (395 and 982 people per km^2, respectively)—as well as small island states which all lack the territorial resources for anything like an even-handed approach to this question. Therefore, all of the world's countries would necessarily have to be divided up into two or perhaps even three categories.

The first category would include states whose territories and demographic density can make it possible to free their respective natural ecosystems from anthropogenic pressure without any significant harm to the population: Russia, Canada, perhaps the United States, the countries of the Scandinavian Peninsula, a significant number of the developing Latin American nations, etc. These so-called donors of ecological space would eventually become the base or bridgehead of biospheric stabilization. Since the whole world would be benefitting from this effort, the other countries can be viewed as ecological consumers or recipients. Some countries may belong to neither group (e.g., China, South Africa, and others).

However, this is only one aspect of the solution. The problem is that the world's main financial and economic potential is concentrated mainly outside of regions with a few islands, islets, and oases of intact wild nature. In this connection, many recipient countries could act as *financial donors* that would subsidize the regeneration of natural ecosystems in other parts of the world. This would pertain particularly to developed countries—the main polluters and emitters of greenhouse gases. Today they exist and prosper essentially by "renting" the remaining intact ecological spaces of other regions.

Unfortunately for ecological donor countries, this rent has not yet been grasped conceptually at an international level, but this ambivalent state is not likely to last. Furthermore, we still don't have a clear notion of how much it really costs to use or possess what we refer to as ecological space. At any rate, this cost is not what urban dwellers bear when it comes to the equipment and services associated with water delivery to their taps. The same applies to farmers and property owners with respect to the real cost of the land they require. After all, land is priced not so much in connection to the needs of natural communities inhabiting a given territory, as it is in terms of the "services" that these life-sustaining communities can supply. These services are normally of a limited (and at the same time very specific) nature (e.g., the natural qualities of a recreational forest with a potential for resort development). Still, what counts more often is the fertility of the soil with agricultural prospects, as well as the location of certain lands intended for urban or industrial development. And finally, the wealth of resources underground enters the equation. Therefore, when it comes to determining the value of a given territory, considerations of the role played by such an area as an element of the biosphere in the maintenance of regional or planetary stability rarely win out in the face of short-term prospects for financial gain.

In the former U.S.S.R. there was once an attempt to rely on expert assessment in order to determine the monetary value of the resource-saving and resource-regenerating role played by wildlife conservation areas (i.e., the value of their contribution to biospheric stabilization). The experts argued that this amounted to 2,000 rubles/ha (about \$500/ha in today's terms; Losev *et al.*, 1993). However, in the year 2000, gross output was of the order of \$30 trillion to \$32 trillion for 86 million km^2 of destroyed ecosystems on land. Therefore, 1 ha of deformed or destroyed ecosystems made a contribution to the world's gross output equaling approximately \$4,000. When this is compared with the above-cited Soviet estimate, the difference is a factor of 8! And so, any economist unburdened by ecological scruples can easily prove the unprofitability of preserving natural ecosystems. However, this unprofitability pertains only to the present moment, and one hardly needs to be told yet again that clean river water and undisturbed forests have no price tag. When choices have to be made regarding the use of our resources for such gargantuan undertakings as manned travel to other planets or yet another dazzling Olympic spectacle, human ambition and megalomania may be too expensive in more ways than one. This is where national pride can be a truly destructive impulse.

In order to give some concrete shape to the above discussion, let us turn to an example from Russia. This country is of particular interest in this instance because its geographic position and natural features give it a key role on the global ecological stage. Like any other industrialized nation, Russia also contributes to the pollution and distortion of the environment, including the emission of anthropogenic carbon. In the year 2000, these emissions reached $0.54\,Gt\ C\,yr^{-1}$ while its share in the global emissions of anthropogenic CO_2 equaled 5.7%. Out of the $0.54\,Gt\ C\,r^{-1}$, if atmospheric mixing is taken into account, $0.31\,Gt\ C\,yr^{-1}$ is absorbed by ecosystems and the waters of the World Ocean. At the same time, the runoff of atmospheric CO_2 into Russia's own forest ecosystems is $0.33\,Gt\ C\,yr^{-1}$ to $0.67\,Gt\ C\,yr^{-1}$ according to various estimates. Total CO_2 runoff into all intact ecosystems on Russian territory reaches $1.0\,Gt\ C\,yr^{-1}$ (Losev, 2001). Therefore, not only is Russia not a donor of atmospheric CO_2, but it also possesses (in comparison with other countries) an unused potential for CO_2 emissions estimated at $0.3\,Gt\ C\,yr^{-1}$ (Zalikhanov *et al.*, 2006). It is consequently clear that this gap is filled by other industrialized states whose territories lack the sufficient natural ecosystems for absorbing their own CO_2 emissions. In other words, Russian ecological space is being "used" for free.

Let us now recall what we said in Chapter 12 about the anti-ecological turn of Russia's state policy and the associated dangers for the country's natural resources that are increasingly under pressure from economic forces. Already today, in the heart of Siberia we find the Krasnoiarskaia, Soiano-Shushenskaia, and Nizhne-Kureiskaia hydroelectric power stations. The next decade will see the construction of the enormous Evenkiiskaia station with a capacity of nearly $8\,GW_t$, a dam around 200 meters high and a flooding area in the order of a million ha (Nefiodov, n.d.). And this is happening on a territory recognized by UNESCO as the cleanest on the planet! Here is what the *Atlas of Russia's Intact Forest Landscapes* has to say on the subject:

> "The Russian forest is no longer a boundless belt of unbroken wilderness. It is better described as a belt of intact fragments that are separated from each other by areas affected either by land use or its side effects ... In Western Siberia, the northern parts of Eastern Siberia and the Far East, the major causes of fragmentation and disturbance are extraction of mineral resources (including prospecting and construction of transportation infrastructure) and the massive human-induced fires which accompany these activities" (Aksenov *et al.*, 2002).

And so, instead of considering the possibility of vacating the sparsely populated regions of Siberia and the far north (which would be the right thing to do under current circumstances), there are preparations being made for the further reclamation of this territory (i.e., the continuation of civilization's assault on nature). Meanwhile, considerations of sustainable development for Russia and the world in general point in the opposite direction (i.e., the concentration of the country's economic potential on its historically reclaimed areas) which would correspond to approximately 30% of Russia's entire territory. If such a measure were to be agreed upon with the entire international community, a significant contribution would be made toward stabilization of the global environment and the halting of atmospheric CO_2 accumulation. This would be Russia's share in the general transition to sustainable development if the concept were to be applied in its scientific sense.

Now, is 30% of Russia's territory a lot or a little, and would the above-mentioned displacement be associated with any kind of sacrifices on the part of the country's population? We are talking about Russia's European territory west of the Ural Mountains, the Ural region south of the 60th parallel and the southern edge of Siberia and the Far East. This area is 5 million km^2, and here we find the concentration of over 95% of the country's industrial potential, as well as 100% of its agricultural land. So no major dislocation is being proposed here. As far as natural and climatic conditions are concerned, this territory happens to be very favorable for human habitation. Furthermore, rendering the economic infrastructure more compact would significantly raise the efficiency of Russia's economy and eliminate the overextension of its lines of communication and roads. This would end up reducing unjustifiable expenditures for energy and transportation. Such a development plan essentially corresponds to the country's economic needs and the interests of its population. This would also benefit the indigenous peoples of Russia's north whose present quality of life is very low and who are now practically on the brink of extinction as nationalities or ethnic groups. First, they would have jurisdiction over a clean natural environment, and, second, they would be given the opportunity to regain their cultural roots and traditional economic activities (e.g., trapping and hunting).

The question is how one would acquire the means for such a large-scale project? Given the present position of the Russian administration, one can hardly expect funds to be budgeted for this purpose in the foreseeable future. Therefore, there is no possibility of avoiding investments from the developed nations. At the same time, Russia itself should be interested—in social and economic terms—in the implementation of such a program, realizing the benefits of keeping its natural wealth intact instead of exploiting it in a primitive economic manner. And so, countries that are benefiting from Russia's CO_2 absorption capacity (given the global nature of the problem) today should be regularly paying Russia the equivalent of ecological rent for preserving the areas in question. According to international experts, the assumed yearly expenditures on the part of the U.S.A., Japan, and West European countries related to fulfilling their obligations for the reduction of carbon emissions amount to $550 to more than $1,100 dollars per ton of carbon (Bedritsky, 2000). Therefore, the ecosystems of Russia—even on the basis of minimal estimates—yearly take out of the atmosphere $160 billion to $325 billion worth of accumulated anthropogenic carbon. This is the amount that essentially Russia "invests" in the international community, and most importantly—in the developing countries.

Sources and runoff channels for carbon have also been estimated with respect to other regions on the continental scale—Asia (without the former Soviet States), Africa, Western and Central Europe, both Americas. and Australia (Kondratyev et al., 2004). As we have already pointed out, Asia is a major climate destabilizer on the planet. It is followed by North America and Europe—primarily because of industrial emissions. So far, Africa is close to neutral while Australia and especially South America remain territories that stabilize the climate—thanks to the intact ecosystems at their disposal (see Tables 15.1 and 15.2). Unfortunately, the fact that the Kyoto Protocol concentrates on CO_2 emissions from fossil fuels and ignores the land use factor, as well as the runoff of anthropogenic carbon into intact ecosystems, creates favorable conditions for some countries and unfavorable ones for others. Meanwhile, in the interests of stabilizing the global environment we need to fundamentally reconsider our attitudes toward rainforests and the forests of the taiga, as well as the associated investments, which could outweigh the material gain from deforestation in Russia as well as countries in the tropical belt. The politicians must change their way of thinking and start relying on science, which has already established biospheric laws. Today, these laws have been dangerously violated, which has brought us to a state of war with life on Earth. And by definition, humanity cannot

Table 15.1. Global sources and sinks of carbon for the period 1991–1994 (Gt/yr) (from Kondratyev et al., 2004, table 8.6).

Sources	Sinks
Fossil fuel burning (5.9)	Land ecosystems (2.5)
Land use (6.3)	Physico-chemical ocean system (2.6) Atmosphere (2.2) Biological pump of the ocean (4.9)
Total = 12.2	Total = 12.2

Table 15.2. Sink of carbon for forests of Canada, Russia, and the U.S. (from Kondratyev *et al.*, 2004, table 8.7)

Country	Satellite data			Data from other sources		
	Pool (Gt C)	*Sink* (Gt C/yr)	*Area* (Mha)	*Pool* (Gt C)	*Sink* (Gt C/yr)	*Area* (Mha)
Canada	10.56	0.0731	239.5	11.89	0.093 0.085	244.6
Russia	24.39	0.2836	642.2	32.86	0.429 0.058	816.5 763.5
U.S.A.	12.48	0.1415	215.5	13.85	0.167 0.098 0.020	217.3 247.0

win since we happen to be on both sides of the conflict at the same time. In the materials of the 2001 *Global Change Open Science Conference* in Amsterdam on the fulfillment and prospects of the main global scientific programs, we find that the Earth is a system where life itself helps to control the state of the planet. Biological processes actively interact with the physical and chemical processes in forming environmental characteristics. However, biology plays a far more important role in maintaining environmental habitability than we previously supposed (Steffen and Tyson, 2001).

In the 18th and 19th centuries the Earth's population was increasing mainly in Europe, which utilized emigration to the Americas and Australia as a valve to let out the demographic "steam". Following Columbus' four voyages and Magellan's circumnavigation of the planet, the European globalization drive began to gather momentum. This unprecedented territorial expansion went on for nearly four centuries, moving the world's borders virtually to the two poles. Although this occurred at the price of the injustices associated with colonialism and imperialism, in certain parts of the world the technological and scientific advances of European civilization led to the reduction of hunger, improvement of health conditions. and a resulting population growth. Therefore, European expansion is partly responsible for the demographic explosion and the associated rapid deterioration of the Third World's remaining ecosystems, whose territory shrank from 80% to 40% of the total land area in the 20th century alone. The realization that this process occurred has been very recent, and only now has it become clear that we need to pay for past mistakes.

When war veterans recall their battle experience, they tend to agree that the greatest trial of armed conflict is retreat—even when only a regiment or a division has to pull back, not to mention an entire army. Special disciplines and mindsets are required to make it work. Similarly, the retreat of a people in the face of an invasion is viewed as one of the most tragic events in its history. What can we say then about the retreat of an entire civilization (i.e., all of modern humanity)? No one is saying that reassessing our fundamental values and acting upon this reassessment would be easy, however unavoidable this may be. This would surely involve reconsidering what qualities we expect in our leadership. To put it bluntly, the world needs Kutuzovs instead of Napoleons and Roosevelts or Erhardts instead of Alexander the Greats. Equally necessary are new Leo Tolstoys who could grasp and convey the fundamentally new reality in artistic terms, providing impetus for change and resolve in a difficult time. This was, incidentally, the case with *War and Peace*, which gave so much strength to millions of Russian people in the darkest hours of World War II. These and many other ideological changes are necessary before humanity finally sees the light at the end of the proverbial tunnel.

To conclude this chapter, we provide Table 15.3 reflecting the trends in environmental change that took place within a period of 20 years, as well as forecasts up to the year 2030 made 15 years ago by a group of Russia's leading ecologists. Unfortunately, we are already witnessing the gradual realization of these predictions, which should parry any possible accusations of alarmism leveled at this book.

Table 15.3. Environmental change 1972–1992 and predicted trends up to 2030 (Losev *et al.*, 1993; Danilov-Danil'yan *et al.*, 1994; updated).

Processes and events	Trends for 1972–1992	Forecast up to 2030
Reduction of natural ecosystem area	On land per year: 0.5–1.0% By the early 1990s: ±40% remained	Trend unchanged leading to almost complete disappearance on land
Consumption of net primary production	Increase: 40% on land, 25%—globally (for 1985)	Increase: 80–85% on land, 50–60%—globally
Changes in atmospheric greenhouse gas concentration	Yearly growth from tenths of a percent to the first few percent	Rising concentration, acceleration of CO_2 and CH_4 concentration as a result of ecosystem destruction
Depletion of ozone layer and growth of ozone hole over the Antarctic	1–2% per year; hole area increasing	Trend unchanged even after cessation of CFC emissions by the year 2000
Reduction of forest (primarily tropical) area	Rate: from 117 (1980) to 180 ± 20 (1989) thousand km^2 Forest regeneration with respect to above—1:10	Trend unchanged; reduction of tropical forests: from 18 million km^2 (1990) to 9–11 million km^2. Reduction of temperate forest belt area
Desertification	Expansion of desert area (60,000 km^2 per year). Growth of technology-induced desertification and toxic deserts	Trend unchanged given deforestation, climate change, and rising pollution
Land degradation	Increasing erosion (24 billion t/yr), dwindling fertility, accumulation of pollutants, acidification, salinization	Trend unchanged, rising erosion, and pollution, reduction of agricultural lands per capita
Rising ocean levels.	1–2 mm/yr	Trend unchanged with possible acceleration to 7 mm/yr
Natural disasters and technologically mediated accidents	Rising frequency/yr: by 5–7%. Rising damages/yr: by 5–10%. Rising victim numbers/yr: 6–12%	Trend intensified

Processes and events	Trends for 1972–1992	Forecast up to 2030
Species extinctions	Rapid	Trend intensified in proportion to biospheric degradation
Pollutant accumulation in media and organisms. Migration within trophic chains	Growth of pollutant mass and numbers. Growth of environmental radioactivity, "chemical bombs"	Trend unchanged and possibly intensified
Increasing frequency of diseases from environmental pollution and the destruction of human ecological niches, including new disorders	Developing countries: rising poverty, food shortages, rising child mortality and illnesses, lack of clean drinking water. Developed countries: high accident frequency, high medication usage, rising allergy rate, growing world AIDS pandemic, decreasing immunity to some diseases and microorganisms	Trend unchanged, growth of food shortages, growth of disease from ecologically based factors, increasing infectious disease territories, appearance of new illnesses

References

Aksenov, D.; D. Dobrynin; M. Dubinin; A. Egorov; A. Isaev; M. Karpachevskiy; L. Laestadius; P. Potapov; A. Purekhovskiy; S. Turubanova *et al.* (2002). *Atlas of Russia's Intact Forest Landscapes*. Moscow. Available at *http://www.forest.ru/eng/publications/intact/5.html*

Bedritsky, A. I. (2000). "Key questions pertaining to Kyoto Protocol negotiations," in *Ekologicheskie aspekty energeticheskoi strategii kak factor ustoichivogo razvitia*. Moscow: Izdatel'skii dom "Noosfera" [in Russian].

Bryant, D.; D. Nielsen; I. Tangley (1997). *The Last Frontier Forests*. Washington, D.C.: World Resources Institute.

CIA (2004). *World Fact Book*. Washington, D.C.: Central Intelligence Agency. Available at *https://www.cia.gov/library/publications/the-world-factbook/*

Climate in Crisis: The Greenhouse Effect and What We Can Do? (1991). Summertown, Oxford, U.K.: Technology Books.

Danilov-Danil'yan, V. I.; K. S. Losev (2000). *The Ecological Challenge and Sustainable Development*. Moscow: Progress-traditsia [in Russian].

Danilov-Danil'yan, V. I.; V. G. Gorshkov; Yu. M. Arsky; K. S. Losev (1994). *The Environment between the Past and the Future: The World and Russia (An Attempt at an Ecological-*

Economic Analysis). Moscow: VINITI (All-Union Institute of Scientific and Technical Information at the Academy of Sciences of the U.S.S.R.) [in Russian].

Diamond, J. (2005). *Collapse: How Societies Choose to Fail or Succeed.* New York: Viking.

Dolnik, V. T. (1992). "Are there biological mechanisms for regulating human population numbers?" *Priroda* **6**, 3–16 [in Russian]. Available at *http://vivovoco.rsl.ru/VV/PAPERS/ECCE/VV_EH13W.HTM*

Etheridge, D. M.; L. P. Steele; R. I. Langenfels; R. J. Francey; J. M. Barnola; V. I. Morgan (1996). "Natural and anthropogenic changes in atmospheric CO_2 over the last 1000 years from air in Antarctic ice and firn," *Journal of Geophysical Research* **101**, 4115–4128.

Falkowsky, P. G.; A. D. Woodhead (Eds.) (1992). *Primary Productivity and Biogeochemical Cycles in the Sea.* New York: Plenum Press.

Friedli, H.; H. Lotscher; H. Oeschger; U. Siegenthaler; B. Stauffer (1986). "Ice core record of the $^{13}C/^{12}C$ ratio in atmospheric CO_2 in the past two centuries," *Nature* **324**, 237–238.

Gittings, J. (2002). "Growing sex imbalance shocks China," *The Guardian*, May 13.

Global Forest Resources Assessment (2005). "Fifteen key findings." Available at *http://www.fao.org/forestry/foris/data/fra2005/kf/common/GlobalForestA4-ENsmall.pdf*

Gorshkov, V. G. (1980). "Biospheric energy flow and its consumption by human beings," *Izvestia VGO* **112**(5), 411–417 [in Russian].

Gorshkov, V. G.; V. V. Gorshkov; A. M. Makarieva (2000). *Biotic Regulation of the Environment: Key Issues of Global Change.* Chichester, U.K.: Springer/Praxis.

Hannah, L.; D. Lohse; Ch. Hutchinson; J. L. Carr; A. Lankerani (1994). "A preliminary inventory of human disturbance of world ecosystems," *Ambio* **4/5**, 246–250.

Houghton, J. T.; L. G. Meira Filho; B. A. Callander; N. Harris; A. Kattenberg; K. Maskels (Eds.) (1996). *Climate Change 1995: The Science of Climate Change.* Cambridge, U.K.: Cambridge University Press.

Houghton, R. A. (1989). "The long-term flux of carbon to the atmosphere from changes in land use," extended abstracts of papers presented at the *Third International Conference on Analysis and Evaluation of Atmospheric CO_2 Data.* Heidelberg, Germany: WMO University, pp. 80–85

Houghton, R. A. (1993). *Emissions of Carbon from Land-use Change, Snowmass, Colorado, July 18–30.* Global Institute on the Carbon Cycle.

Kondratyev, K. Ya.; K. S. Losev; M. D. Ananicheva; I. V. Chesnokova (2004). *Stability of Life on Earth: Principal Subject of Scientific Research in the 21st Century.* Chichester, U.K.: Springer/Praxis.

Lorius, C.; H. Oeschger (1994). "Paleo-perspectives: Reducing uncertainties in global change?" *Ambio* **3**(1): 30–36.

Losev, K. S. (2001). *Ecological Problems and Prospects for Sustainable Development in Russia in the 21st Century.* Moscow: Kosmosinform [in Russian].

Losev, K. S.; V. G. Gorshkov; K. Ya. Kondratyev; V. M. Kotlyakov; M. Ch. Zalikhanov; V. I. Danilov-Danil'yan; G. N. Golubev; I. T. Gavrilov; V. S. Reviatkin; V. F. Grakovich (1993). *The Ecological Problems of Russia.* Moscow: VINITI (All-Union Institute of Scientific and Technical Information at the Academy of Sciences of the U.S.S.R.) [in Russian].

Lurin, B.; I. Rasool; W. Cramer; B. Moore III (1994). "Global terrestrial net primary productivity," *Global Change Newsletter* **19** (September), 6–8.

Nefiodov, A. V. (n.d.) *Did Russia Have Any Rivers?* Available at *http://www.biotic regulation.ru:80/life/life10_r.php*

Petit, J. R.; J. Jouzel; D. Raynaud; N. I. Barkov; J.-M. Barnola; I. Basile; M. Bender; J. Chappellaz; M. Davisk; G. Delaygue *et al.* (1999). "Climate and atmospheric history of the past 420,000 years from the Vostok ice core, Antarctica," *Nature* **399**, 429–436.

Raynaud, D.; J. Jouzel; J. M. Barnola; J. Chapellaz; D. J. Delmas; C. Lorius (1993). "The ice record of greenhouse gases," *Science* **59**, 926–934.

Rodoman, B. B. (2004). *Russia as an Administrative and Territorial Monster.* Available at *www.polit.ru/lectures/2004/11/04/rodoman.htm*

Rotty, R. M. (1983). "Distribution of and changes in industrial carbon dioxide production," *Journal of Geophysics Research* **88**(C2), 1301–1308.

Siegenthaler, U.; H. Oeschger (1987). "Biospheric CO_2 emission during the past 200 years reconstructed by deconvolution of ice core data," *Tellus* **39B**(1/2), 140–154.

Siegenthaler, U.; J. L. Sarmiento (1993). "Atmospheric carbon dioxide and the ocean," *Nature* **365**, 119–125.

Staffelbach, T.; B. Stauffer; A. Sigg; H. Oeschger (1991). "CO_2 measurements from polar ice cores: More data from different sites," *Tellus* **43B**(2), 91–96.

Steffen, W.; P. Tyson (Eds.) (2001). *Global Change and the Earth System: A Planet under Pressure.* Stockholdm, IGBP Science, p. 4.

UNDP India (n.d.). See *http://www.undp.org.in/index.php?option=com_content&task=view&id=157&Itemid=99999999*

Vitousek, P. M.; P. R. Erlich; A. H. E. Erlich; P. A. Matson (1986). "Human appropriation of the product of photosynthesis," *Bioscience* **36**(5), 368–375.

Vompersky, S. E. (1994). "The biospheric significance of swamps in the carbon cycle," *Priroda* **7**, 44–55 [in Russian].

Watts, J. A. (1982). "The carbon dioxide questions: Data sampler," in W. C. Clark (Ed.), *Carbon Dioxide Review.* New York: Clarendon Press.

Wofsy, S. C.; M. L. Goulden; J. W. Munger; S.-M. Fan; P. S. Bakwin; B. C. Daube; S. L. Bassow; F. A. Bazzaz (1993). "Net exchange of CO_2 in a mid-latitude forest," *Science* **260**(5112), 1314–1317.

World Resources, 1990–1991 (1990). New York: Basic Books.

Worldwatch Database (2000). Washington, D.C.: Worldwatch Institute.

Zalikhanov, M. Ch.; K. S. Losev; A. M. Shelekhov (2006). "Natural ecosystems as the key natural resource of humanity," *Vestnik rossiiskoi akademii nauk* **76**(7), 612–614 [in Russian].

16

Navigation directions: Indicators of sustainable development

Indicators of sustainable development according to the U.N.'s Human Development Index—Evaluation of material flow and the consumption of net primary production in Germany and Austria—Biotic regulation of the environment and priorities in a system of ecological indicators—Hierarchic superiority of ecosystem indicators over environmental pollution indicators—Resource crisis vs. ecological crisis: similarities and differences—Geographic information systems

When we wander over icy terrain, deserts, or uninhabited virgin steppes, our main concern is to avoid losing our way or our sense of present location. In this case maps are very handy, or in their absence one can navigate by the Sun or the stars. Keeping track of every noticeable landmark—a stick, a hill, a dry riverbed, isolated trees—is another tried and tested method. No such helpful tracking clues are available as we meander through the new and unfamiliar realm of sustainable development where the *fata morgana* of wishful thinking constantly threatens to lead us astray. This is why a special section of Agenda 21—the document adopted at the 1992 Rio Summit—was devoted to the question of creating special sustainable development indicators for the systemic evaluation of the relationship between nature, the economy, and population. At the same time, these indicators were meant to act as a basis for working out long-term policies in the framework of corresponding programs. In Chapter 40 of this document we read: "Indicators of sustainable development need to be developed to provide solid bases for decision-making at all levels and to contribute to a self-regulating sustainability of integrated environment and development systems" (*Agenda 21*, 2004).

One of the first projects related to these sustainable development indicators was proposed in 1996 by the U.N. Commission on Sustainable Development (CSD). In this document, 134 indicators were divided into four groups—each reflecting an aspect of sustainable development: social, ecological, economic. and institutional. Here is the general outline:

(1) Social indicators include demographic dynamics, the quality of amenities and facilities in places of human habitation, level of education, health and safety standards (e.g., longevity prospects at birth, child mortality up to age 5), etc.
(2) Ecological indicators include protection of the atmosphere (indicators of greenhouse gas emissions), measures for the prevention of desertification and deforestation, rational land resource use, the maintenance of biodiversity, ecologically safe management of solid radioactive waste and sewage, etc.
(3) Economic indicators include financial resources and mechanisms, per capita GDP index, consumption patterns (including energy use), cooperation in the transfer of environmentally sound technologies, etc.
(4) Institutional indicators—intended to reflect, on the one hand, the state of policies on sustainable development and, on the other hand, the involvement of key population strata in this process—include such issues as the presence of national programs for a transition to sustainable development, the implementation of global accords on sustainable development, the numbers of Internet users per thousand people, etc.

The document describes the applicability of every indicator, methods for their calculation, their relationship to sustainable or non-sustainable development, and a series of other parameters (U.N., 1996).

Obviously, these indicators were not included merely for informational purposes, but rather primarily as a means of achieving or adjusting sustainable development. In this practical context, the project required a great deal of work for narrowing down and clarifying terminology, coordinating the indicators with Agenda 21, as well as working them out in quantitative terms (which was by no means possible in all cases). Furthermore, the use of numerous indicators proposed by the U.N. project raised difficulties in certain developing countries because the required information and statistical data were lacking.

At the same time, certain indicators in the document (e.g., literacy level) were not relevant to developed countries. Therefore, the package of indicators did not apply in its entirety to most countries in light of their real circumstances and specific characteristics. For example, in their respective sustainable development reports for 1997, Finland included only 57 out of 134 indicators, the U.K. 41 and the U.S.A. 56, of which most were qualitative. However, what matters in this case is that the notion of a unified approach to the assessment of progress—or, conversely, regression—on the way to sustainable development, by means of comparable quantitative and qualitative criteria, turned out to be quite fruitful and appealing to most countries. To date, a substantial amount of constructive experience has been accumulated in this area: for example, valuable results have been obtained with respect to adapting various indicators to regions of different sizes—from entire states to individual cities or even small districts.

Special mention ought to be made of the so-called aggregate indicators of sustainable development proposed in different years independently of the U.N. project. These indicators allow for a complex picture of the state in which a given country or region finds itself, and what trends of development can be expected. For example, in 1990 the Pakistani economist Mahbub ul-Haq created the Human Development Index (HDI) also known as the *Quality of Life Index*. The latter is derived from a calculation of the expected longevity index at birth, the literacy and education index (the percentage of a given adult population with primary, secondary, and higher education), as well as the GDP index. For example, the longevity index is calculated as follows: expected longevity at a given point in time in a given country minus the minimal world longevity average, divided by the difference between the maximal and minimal world longevity values. The GDP index is derived through a somewhat more complex calculation; however, here too the basis consists of average world GDP indicators, as well as the minimal and maximal GDP per capita. In this case, it is assumed that an index below 0.5 characterizes countries with poor human development while indices of 0.8 and up are indicative of high human development.

Of course the HDI is not an all-inclusive indicator of the quality of life and human development; however, it does reflect this characteristic far more fully than, for example, the GDP and communicates a more complex (three-dimensional) picture of the way human society is progressing, making it possible in particular to note the interconnection between the quality of life and material well-being. As of 1990 the U.N. has been publishing *Human Development Reports* as official yearly documents that make it possible to rank the world's countries in accordance with the HDI and establish the dynamics applicable in this context. These reports demonstrate that for most countries, the HDI has been experiencing slow but constant yearly growth so that fewer and fewer states appear in the low HDI category. The latter point, unfortunately, does not apply to sub-Saharan Africa. Extreme economic backwardness combined with a low level of education and high mortality rates made worse by the AIDS epidemic firmly keep the peoples of Africa in a vice, depriving them of the chance for a dignified human existence. For instance, the life expectancy of the average inhabitant of Niger (a state occupying one of the bottom rungs in the table featured by the *Human Development Report* for 2007–2008) is expected to be half as long as that of the average Norwegian. The yearly income discrepancy between these two extremes is over 40 times in favor of the Norwegian, and the citizen of Niger will suffer from severely limited access to education; all levels of education are available to only 21% of the population.

From a total of 177 countries for which the 2007–2008 report provides calculations (using 2005 data), 22 have a low level of HDI and are all in Africa. Another 85 "Third World" countries—which also include China that is located in 81st place in the general rating, as well as certain former Soviet republics—constitute a group with an intermediate level of HDI. And finally, 70 countries are rated at a high level of HDI. These include first of all the developed states of Europe, North America, Australia, Japan, and New Zealand, as well as most Latin American countries and Russia—which occupies rung number 67. The first ten places in this group are distributed as follows: Iceland and Norway are on top with an index of 0.968, followed by Australia, Canada, Ireland, Sweden, Switzerland, Japan, The Nether-lands, and France (UNDP, n.d.).

Another extremely important aggregate index of sustainable development is the evaluation of material (physical) flow in individual countries. Both the economic and ecological well-being of any country are directly related to this form of flow. When the latter is diminished through the implementation of energy-saving technologies, local ecological problems and to a certain extent global ones can be solved (cf. Meadows et al., 1992; Weizsaecker et al., 1997). One of the first attempts to evaluate material flow took place in Germany in the early 1990s. Based on the resulting balance of substance consumption, it was determined that the yearly per capita use of raw materials was almost 75 t. Of this amount, 56 t were waste and 19 t—products, including construction objects (which essentially amounts to delayed waste). Furthermore, three-fourths of the material flow in Germany corresponded to imports, which in absolute numbers amounted to 433 million t. The latter in turn constituted 2.1 billion t of waste in countries that had provided the raw materials to Germany, which implies 304 tons of soil erosion in these exporter countries.

As for local substance flow, it included 3.99 billion t of abiotic raw materials and 82 million t of biotic materials (as dry substance). After use, the abiotic raw materials—minerals, metallic ores, and energy sources—produced 2.89 billion t of solid waste. Of that amount 222 million t ended up in controlled waste deposit sites while 2.7 billion t were buried in mine shafts or in the soil. The yearly greenhouse gas and pollutant emission rate was 1.6 billion t with an additional 34 million t of such substances ending up in sewage waters. The creation of biotic raw materials on Germany's fields came at the cost of soil erosion amounting to 129 million t. The processing of these raw materials into products required the use of 1,070 billion t of air and 70 billion t of water. And so, it turned out that Germany's economy grew materially by almost 1 billion t/yr—mainly through the accretion of material required for the construction of roads, buildings, machines, and mechanisms—while the vol-ume of per capita substance transfer was 1.5 times above the world average (the latter being 50 t). This material physical growth was inexorably eating up the country's natural space (i.e., islets of remaining ecosystems, agricultural land, and forests). Furthermore, Germany's contribution to pressure on the environment in certain other world regions was considerable—about 70% (Bringezu et al. 1996). It is more than clear that such development can hardly be called "sustainable".

While the estimation of material flow is quite labor-intensive and requires com-plex calculations, as well as a great deal of research, the other aggregate index of sustainable development (energy use or the utilization of energy capacity per unit of territory) is much more universal and does not involve large-scale data collection and calculations. In order to create something in the economic sphere, people need lighting, heated (or cooled) facilities, machines and mechanisms, transportation and communication, as well as computers (as of the past two decades or so). All this requires materials, space, and energy, which means that increases in any sphere of activity are inevitably accompanied by the destruction of the natural environment and its replacement by such artificial entities as agricultural land, residential and industrial facilities, transportation infrastructure, etc. Therefore, any form of energy use implies the same thing for the environment: its distortion or destruction. The overall yearly volume of energy use by man for this "war on nature" can easily be compared with the energy released by a nuclear war—only in this instance the energy is dispersed in time and space (Danilov-Danil'yan and Losev, 2000). This is why human pressure on nature can be viewed in terms of energy investment per unit of reclaimed territory, and there are a number of examples of how this integral indicator is used in the specialized literature (cf. Danilov-Danil'yan *et al.*, 1994; Kotlyakov *et al.*, 1995; Arsky *et al.*, 1997).

The net primary production index occupies a special position among indicators of sustainable development for a given territory. Unfortunately, the index of material flow is insufficient to describe the interaction between nature and society in a relevant way. As it maintains its own "metabolism", society intrudes into natural ecosystems and radically denatures them. An example of this so-called "colonization of natural systems" would be the way agriculture not only alters biotopes but also their resident organisms—right down to the genetic level. Industry, on the other hand, simply destroys natural ecosystems instead of merely changing them. Therefore, the scale on which humans consume net primary production (the basis of life on Earth and the source of nourishment for consumer-organisms) can be viewed as an important generalizing indicator for the extent of this intrusion.

As has been mentioned above, the consumption of net primary production is possible in two forms:

* its direct appropriation or "seizure" (e.g., through the utilization of wood);
* the diminution of its productivity by the displacement of natural biota (e.g., the replacement of natural ecosystems by less productive agrocenoses or the construction of roads, buildings, and facilities, as well as the formation of waste sites and vacant lots, etc.).

Net primary production can be quantified in units of mass for organic carbon (tons), power (watts), or energy (joules)—corresponding to the volume of biomass produced by plants on a given territory per year, minus expenditure for respiration and the growth of the plants themselves.

The consumption of net primary production in Austria was evaluated in 1991. The overall energy of plant cover over the country's territory under natural conditions was estimated at 1.501 PJ/yr while its physical diminution according to various categories was as follows:[1]

* land reclamation for construction—7%;
* land reclamation for agriculture—21%;
* land reclamation for forestry—14%.

In this manner various types of consumption in Austria amounted to the yearly removal of 42% of net primary production, of which 21% occurred through direct

---

[1] 1 PJ = 1 petajoule = 10^{15} joules.

transfer into the anthropogenic channel (Haber, 1995). This is slightly above the world average, which equals 40% (Gorshkov, 1980; Vitousek *et al.*, 1986) although somewhat lower than in most European countries (not counting Russia and perhaps Scandinavia) where more territory (than in Austria) is taken up by developed lands, agricultural land, and artificially forested areas that have almost entirely replaced natural forests. As a final note, Austrian ecologists have expressed the opinion that the net primary production index should occupy first position among the indicators of sustainable development.

Although ecological indicators are usually given a certain priority in most proposed projects, it is at times difficult to trace any clear pattern of preference within this group itself. For example, in the above-mentioned project worked out by the U.N. Commission on Sustainable Development (CSD), ecological indicators are divided into 14 groups, most of which have to do with pollution prevention, the rational handling of toxic or dangerous waste, the use of biotechnologies, etc. However, there is not a single indicator pertaining to the human consumption of net primary production, and an indicator as important as the state of ecosystems is mentioned only in connection with the vulnerable ecosystems of mountains and arid regions characterized by low levels of productivity. Equally absent are indicators for the degree of disruption of forest ecosystems (if one does not count the correlation between natural plantation forests and the portion of protected forests for an entire forested area). Finally, the group of indicators regarding the preservation of biodiversity is limited to only the percentage of disappearing species and the portion of specially protected areas including recreational lands and cultural heritage sites. And yet, it has long been clear that the preservation of biodiversity is impossible without maintenance (on an appropriate scale) of all natural ecosystems types. Furthermore, specially protected territories are limited in area and therefore cannot offer on their own a solution to this problem.

Therefore, it would not be an exaggeration to say that without some kind of systemic basis, the ecological indicators in this and other projects end up forming a mosaic that fails to reflect the real state of the environment in its most critical aspects. This is why the issues at hand should be tackled based on the theory of biotic regulation—in particular with respect to the biosphere's carrying capacity as an integral indicator of the uppermost allowable human impact on the biosphere. In light of the above, the highest priority should be accorded to indicators that correlate directly with the carrying capacity. These are first of all indicators of natural eco-systems that ought to reflect the following:

(1) the portion of a given country's or region's entire territory occupied by natural ecosystems;
(2) the size of territory which natural ecosystems should occupy in order to ensure effective environmental regulation;
(3) the rate at which natural ecosystems are reduced or regenerated.

All of these indicators are fully capable of characterizing both the current ecological situation on a given territory and a vector of ongoing change. In other words, we are concerned with how close a country can approach ecological sustainability.

Of equal importance are indicators pertaining to the human consumption of net primary production. They point to the following:

(1) the actual state of a given territory;
(2) the relationship between the actual consumption of net primary production and the allowable limit of its consumption for various ecosystems (in relative and absolute terms);
(3) the rate and vector of change within a given time interval.

In a way, these two groups of leading indicators reflect such key aspects of sustainable development as the question of greenhouse gases, as well as the protection of the World Ocean, seas, and shorelines; the preservation of vulnerable ecosystems and the prevention of deforestation; the safety of freshwater resources; and the problem of biodiversity. Finally, they represent an integrated approach to the planning and management of land resources.

When compared with the above two categories, environmental pollution indicators should be assigned a lower position in the priority hierarchy. They primarily reflect local processes while ecosystem indicators and those pertaining to the consumption of net primary production have to do with local, as well as global changes.

Similarly, in the framework of the same approach we can solve the question of indicators pertaining to social, economic, and institutional aspects of sustainable development. These must also be integrated with the key strategic goal of sustainable development: namely, to direct the development of civilization toward real sustainability commensurate with the laws of the biosphere and biospherically imposed limits by signaling whether we are moving away or toward the corresponding benchmarks. For example, among the economic indicators, priority should be given to those allowing us to prevent the further destruction of natural ecosystems on the one hand, and, on the other hand, those facilitating the territorial expansion of natural ecosystems. This is particularly relevant for virgin forests, wetlands, and the ecosystems of the World Ocean. In particular, the indicators having to do with effective resource use should be constantly directing the economy toward the reduction of anthropogenic pressure on existing natural ecosystems. This would be facilitated by a shift to resource-saving technologies that make it possible, among other things, to buy time for the solution of humanity's main strategic goal: the regeneration of destroyed ecosystems.

In this connection, we must not forget about the fundamental difference between a resource crisis and an ecological crisis since these concepts are often confused (cf. Meadows *et al.*, 1992; Liuri, 1997). This distinction is extremely important not only in theoretical terms but also with respect to practical policies. For example, we refer to a resource crisis in connection with the depletion of some raw material or energy source. Normally, such a crisis can be overcome through the creation of new technologies (i.e., the exclusion of a given resource from the technological cycle or its adequate replacement). An ecological crisis is a phenomenon of a fundamentally different order related to the depletion of natural mechanisms for ensuring environmental stability. And so, in the latter case a solution must be sought from an entirely different direction. Essentially, for the past hundred years, our civilization has already been successfully (as some have thought) evolving under conditions of a full-scale ecological crisis that began long before we reached any sort of resource limits (Gorshkov, 1995). However, the signals of this crisis are not being perceived fully or end up being interpreted in terms of resource depletion. That is why a correctly chosen system of priorities is of such importance—also in connection to indicators of sustainable development.

Finally we have to address the question of computerized geographic information systems (GISs) that are invaluable when it comes to creating a concrete spatial understanding of ecologically significant biospheric and social processes. A feature of these systems is the capacity for cartographic visualization of extensive databanks. Indicators based on GIS make it possible to represent not only spatial but also temporal aspects of environmental change (e.g., by creating a graphic picture of how territories with undisrupted ecosystems are shrinking). The Google Earth program available to anyone on the World Wide Web allows people to view the state of natural ecosystems as if from space on a scale ranging from smaller regions and individual countries to entire continents. GIS maps showing the spatial distribution of certain key social and economic indicators—population density and growth, birth coefficients, agricultural productivity, etc.—are now available for whole continents (*Sustainability Indicators*, 1997).

References

Agenda 21: Chapter 40 (2004). New York: United Nations. Available at *http://www.un.org/esa/ sustdev/documents/agenda21/english/agenda21chapter40.htm*

Arsky, Yu. M.; V. I. Danilov-Danilian; M. Ch. Zalikhanov; K. Ya. Kondratyev; V. M. Kotlyakov; K. S. Losev (1997). *Ecological Problems: What is Going On? Who Is To Blame? What Should Be Done?*. Moscow: MNEPU [in Russian].

Bringezu, S.; R. Behrensmeier; H. Schuetz (1996). "Material flow accounts including the environmental pressure of the various sectors of the economy," *International Symposium on International, Environmental and Economic Accounting in Theory and Practice, Tokyo, March 5–8*.

Danilov-Danil'yan, V. I.; K. S. Losev (2000). *The Ecological Challenge and Sustainable Development*. Moscow: Progress-traditsia [in Russian].

Danilov-Danil'yan, V. I.; V. G. Gorshkov; Yu. M. Arsky; K. S. Losev (1994). *The Environment between the Past and the Future: The World and Russia (An Attempt at Ecological-Economic Analysis)*. Moscow: VINITI [in Russian].

Gorshkov, V. G. (1980). "The structure of biospheric energy flow," *Botanical Journal* **6**(11), 1579–1590 [in Russian].

Gorshkov, V. G. (1995). *Physical and Biological Basis of Life Stability: Man, Biota, Environment*. Berlin: Springer-Verlag.

Haber I. H. (1995). "Menschliche Eingriffe in den naturlichen Energieschlass von Oekosystemen, Sociooekonomische Anegnung von Nettoprimaerproduktion on den Bezirken Oesterreichs," *Social Ecology Papers* **43**.

Kotlyakov, V. M.; K. S. Losev; I. A. Suietova (1995). "Energy investment in a given territory as an ecological indicator," *Bulletin of the Russian Academy of Sciences, Geography Series* **3**, 70–75.

Liuri, D. I. (1997). *The Development of Resource Use and Ecological Crises.* Moscow: Delta [in Russian].

Meadows, D. H.; D L. Meadows; J. Randers (1992). *Beyond the Limits: Confronting Global Collapse, Envisioning a Sustainable Future.* Post Mills, VT: Chelsea Green.

Sustainability Indicators (1997). *A Report on the Project on Indicators of Sustainable Development.* Chichester, U.K.: John Wiley & Sons.

U.N. (1996) *Indicators of Sustainable Development: Framework and Methodologies.* New York: United Nations. Available at *http://esl.jrc.it/envind/un_meths/UN_ME001.htm*

UNDP (n.d.) *Human Development Report 2007/2008.* New York: UNDP. Available at *http://hdr.undp.org/en/media/hdr_20072008_en_complete.pdf*

Vitousek, P. M.; P. R. Erlich; A. H. E. Erlich; P. A. Matson (1986). "Human appropriation of the product of photosynthesis," *Bioscience* **36**, 368–373.

Weizsaecker, E. U. von; A. B. Lovins; L. H. Lovins (1997). *Factor Four: Doubling Wealth Having Resource Use (A Report to the Club of Rome).* London: Earthscan.

Part VI

"Is there enough community, responsibility, discipline and love?"
(Meadows et al., 1992)

17

The barricades of old thinking in the way of sustainable development

"Non-scary" aspects of the ecological crisis—Why there is no mention of a global crisis in the Global Trends 2015 report of the American National Intelligence Council—The educational paradigm of sustainable development—Self-fulfilling model of the future—Each man for himself—Can the "golden billion" save itself?

If, as has been argued so far, the situation is so grave, why has the world not managed to implement the policies required to save the biosphere? Why are so many buzz-words, like "Kyoto", "ozone layer", and "oil spills" flying about, so many alarm bells ringing out of every media source, and yet people still go about their daily business of driving, flying, producing, and consuming? Why is faith in a technological solution still so powerful in our progress-oriented society that many people, including decision-makers all over the world, keep clinging to the idea that only some aspects of our life-style need to be changed while the basic structure can remain unaltered?

The notion of sustainable development is simply not reflected in *Global Trends 2015* (NIC, 2002)—the already cited report of the U.S. National Intelligence Council prepared by foremost American experts and aimed (as the title suggests) at global solutions. Here you can find statements about the positive and negative consequences of globalization and the place occupied by the U.S.A. in this process. It is said that "the United States, as a global power, will have little choice but to engage leading actors and confront problems on both sides of the widening economic and digital divides in the world of 2015." *Global Trends 2015* discusses good growth prospects for the world's economy that "will return to the high levels reached in the 1960s and early 1970s, the final years of the post-World War II 'long boom'." At the same time, we read about such challenges as the impending freshwater shortage threatening more than 3 billion people, shrinking tropical rainforests, melting ice caps, rising sea levels, and other "issues" that humanity will have to deal with.

However, *Global Trends 2015* says nothing about environmental deterioration as a truly global or integrated phenomenon that sets limits to economic and demographic growth. Instead the authors of the report tend to view everything in fragmented terms, each problem existing in isolation and requiring an isolated solution. This approach makes everything seem far less urgent than it is, creating a false sense of security and the illusion that the basic economic equation is unchanged.

Global Trends 2015 admits that, "given the promising global economic outlook, greenhouse gas emissions will increase substantially. The depletion of tropical forests and other species-rich habitats, such as wetlands and coral reefs, will exacerbate the historically large losses of biological species now occurring." And yet, the authors argue: "developed countries will continue to manage these local environmental issues [despite the increase in greenhouse gas emissions?] and such issues are unlikely to constitute a major constraint on economic growth or on improving health standards." They assume that only the developing world will "face intensified environmental problems as a result of population growth, economic development, and rapid urbanization." But the conclusion is essentially "business as usual".

It is difficult to imagine that the authors of *Global Trends 2015* are simply ignorant in this matter or unaware of the Club of Rome Reports, the conclusion of the Brundtland Commission, the work of Donella and Dennis Meadows, Peter Vitousek, Herman Daly, and a number of others, including American researchers who have convincingly demonstrated the specifically global nature of the ecological threat extending to both the industrialized and the developing worlds. One possible reason for such an intractable position is the basic conservatism of prevailing ideologies in the post-Communist world. Having won the big contest with the Eastern Block, the Western world appears to be on the path of consolidation—despite the "green" rhetoric of numerous politicians. The *realpolitik* compromises of the German Green Party in the Schröder government from 1998 to 2005 constitute perhaps the most eloquent example of this trend.

Radical change has always been a painful process that people naturally tend to avoid if they can help it. It took three centuries for Christianity to replace paganism in ancient Rome, which finally happened under Emperor Constantine in 313 ACE. Many more centuries passed before the idea of a geocentric universe was given up and replaced with the heliocentric model of Copernicus in the 16th century. The persecution of Galileo as a champion of Copernicanism demonstrates that even in the presence of incontrovertible evidence, resistance to change is a tremendous force not to be underestimated. Admittedly, the rate of social, political, and ideological change has been increasing enormously—especially since the advent of the 20th century. However, the problem is that even this acceleration is not keeping up with the speed of ecological deterioration on Earth. Therefore, more than ever, the weight of responsibility on the shoulders of the educational system is unprecedented.

How does one impress upon young minds that the notion of the biosphere's carrying capacity should be considered in the same terms as the basic laws of physics?

Simply teaching children to love nature is insufficient. They need to acquire the habit of global thinking and start considering their immediate or local environment as part of a large, extremely fragile whole—not in theoretical terms, but in relation to their daily actions. The difficulty of implementing such psychological change is easily seen from the amount of time spent by children and teenagers all over the industrialized world as consumers of energy in pursuit of electronic entertainment. "A new system of biological education," says a letter to the newspaper *Izvestia* by professors and staff of Moscow University's Biology Department "is needed to provide basic training in biology and appeal to the feelings, reason and civil responsibility of all people. This is the only way to save humanity from disasters caused by biological illiteracy, environmental slovenliness and excesses of consumer society" (Gusev *et al.*, 2003). In this connection it is especially crucial to teach biological literacy to non-biologists (i.e. those who opt for non-biological professions but may eventually end up making significant decisions on environmental and biospheric issues).

And so, it is the practical rejection of sustainable development—whatever the current "ecological" rhetoric may be—that must be taken into account by analysts attempting to project possible future scenarios. The most obvious model of this sort, as is pointed out in *Beyond the Limits* (Meadows *et al.*, 1992), is the "business as usual" situation: that is, self-generating and ungovernable economic growth, the continuation (however much it may be slowed down by modern technologies) of environmental deterioration, the prevalence of national smugness and egoism often presented in rather respectable packaging, the dominance of economic criteria over ecological constraints, the underestimation of signs pointing to biospheric destruction, etc.

It is quite possible to imagine that this relatively serene stage will quickly become malignant as soon as signs of impending catastrophe become apparent to the majority. One pays little attention to good manners on a sinking ship, and so many norms of international relations, which took so much time and effort to attain, can be dispensed with in an instant. The next stage may involve not just a totalitarian but an ultra-totalitarian scenario (e.g., a brutal military dictatorship with two or three centers of power—reminiscent of George Orwell's dystopia *Nineteen Eighty-Four*— in economically dominant states that enslave weaker "Third World" countries). This is likely to be characterized by such sinister phenomena as a ruthless struggle for natural resources, as well as merciless social and biological eugenics, the latter being part and parcel of totalitarian government on one's own territory.

Obviously the time has not yet come for open calls in favor of world dictatorship by any given geopolitical force. However, undercurrents of this trend appear to lie barely beneath the surface of current events. Thus, the pre-emptive military strikes by the United States and their allies against so-called rogue states can be interpreted as a dress rehearsal before the upcoming "performance". As A. Panarin writes, "the world is coming back to the salvation question that arose before humanity at the dawn of the Christian era: is only one chosen people going to be saved or is salvation for everybody? Today the role of the chosen people is played by the forces of the 'golden billion' which have arrogated all access to the post-industrial future" (Panarin, 2002).

One can already glimpse an "enclave" position in the notion of exclusive eco-logical stability projected by *Global Trends 2015* for the industrialized world as opposed to ecological deterioration anticipated by the same report with respect to the developing world. Apparently, there exists a hope for the survival of the "golden billion" despite an ecological disaster and the demise of those outside the industria-lized enclave. This viewpoint leads to a failure to view the rejection of sustainable development (at least for a few exclusive countries) as something irreversible or fatal.

What kind of a wall can be built by the privileged enclave to keep the rest of the suffocating world at bay? The Great Wall of China failed to protect the "chosen people" from "barbarians", and this paradigm will undoubtedly be repeated. Once the barrier is breached and a wave of ecological refugees spill over into the indus-trialized world, how will the democratic institutions of the West cope with this challenge? Unable to stop the influx of immigrants, privileged countries will be forced to erect a legal wall between their own citizens and the newcomers according to the familiar Orwellian principle "all animals are equal, but some are more equal than others." As Immanuel Wallerstein writes, "in a twinkling of an eye America may find itself in a situation where the bottom 30 and even 50 percent of its workers will no longer be proper citizens, and so they will be denied voting rights and a reliable access to social welfare. Should this happen, we will have to put our clocks back 150–200 years" (Wallerstein, 1995).

However, even the sternest segregation policy can merely prolong the agony. Eventually, as the struggle intensifies and moral values deteriorate in the inevitable rat race, any thought of preservation or long-term planning will become less and less likely. Then, what is to stop the inhabitants of a world out of control—and especially those in the dying "Third World"—from doing away with whatever may be left of the natural environment? These questions come to mind when one tries to imagine attempts to survive rather than prevent an ecological catastrophe. Even if someone does somehow succeed in reaching the last island of civilization, such a person is likely to envy the dead.

References

Gusev, M.V.; V. A. Golichenkov; K. N. Timofeev *et al.* (2003). "Biology courses should be not cut but expanded," *Izvestia*, September 6 [in Russian].
Global Trends 2015, see NIC (2000).
Meadows, D. H.; D. L. Meadows; J. Randers (1992). *Beyond the Limits: Confronting Global Collapse, Envisioning a Sustainable Future*. Post Mills, VT: Chelsea Green, p. 234.
NIC (2000). *Global Trends 2015: A Dialogue about the Future with Nongovernment Experts.* Washington, D.C.: National Intelligence Council, December. Available at *http://infowar.-net/cia/publications/globaltrends2015/*
Panarin, A. S. (2002). *Tempted by Globalism*. Moscow: Russky natsionalnyi fond [in Russian].
Wallerstein, I. (1995). *After Liberalism*. New York: The New Press.

18

What the market economy can and cannot accomplish

Constructive and destructive potential of the market economy—Unregulated development in light of the ecological menace—Overproduction of civilization: external crises and internal crisis—Precedents of state-regulated markets—Can ecology be someone's "internal matter"?—Planetary organization for sustainable development responsible to the biosphere

The logic of the foregoing once again suggests that there is no rational alternative to sustainable development (or whatever term we might use for this concept). This sole acceptable path of humanity into the future, as opposed to inertia, can be called the "transformation scenario". The key features of this scenario are

- Quick realization of the threats associated with environmental destruction.
- Adequate reaction to the socio-ecological crisis.
- A breakthrough toward a new worldview and set of values.
- A broad international partnership based on cooperation and financial aid to developing countries.

It hardly needs saying that this path will be full of thorns, requiring people to be responsible, tolerant, compassionate, and capable of self-limitation: that is, all that the authors of *Beyond the Limits* (Meadows *et al.*, 1992) classify as human virtues. However, it will likely require still more than that: namely, the reassessment of certain principles characterizing modern political culture.

Until now, the development of sovereign states has been mainly natural and unregulated, which by the way prompted the criticism of the capitalist system by Karl Marx and his followers. Proposing to replace this system with a rationally organized, trans-national classless society of the future, they imagined a system governed by a single plan and coordinated from one intellectual center. History has demonstrated the groundlessness and the dangerous flipside of this voluntaristic project compared with which the market economy has turned out to be more flexible, humane, and in the final analysis—more viable. The definitive failure of the Soviet experiment seems to have closed all debate regarding the advantages of centralized methods of government.

Indeed, at the end of the more than 70-year-long contest, the market system scored what was probably its most convincing victory.[1] However, can one consider this system invulnerable and beyond this specific historical confrontation? And does such a victory offer some sorts of "indulgences" for the future? Alas, there are far more questions here than answers, and the main questions naturally revolve around the problem of the ecological challenge. After all, one can hardly deny that it is the market economy that has made the biggest contribution to the current global crisis.[2] Admittedly, it is also this system and its institutions that were the first to recognize the ecological threat, placing the question of sustainable development on the world's agenda. But here the enlightened system begins to spin its wheels instead of moving forward. It turns out that there is a world of difference between outlining the problem and actually implementing solutions.

The practical steps in question, as is becoming more and more apparent, are extremely difficult for the market system not only for subjective reasons, but also for objective ones. Essentially, the system is being asked in a way to jump beyond itself and transcend certain foundations of its almost 400-year-old existence. Can it rise above corporate and narrow national interests, or above market egoism and the pursuit of immediate gain? The shift to sustainable development requires this very transformation—a new way of "breathing" and strategic temporal thinking. However, it appears that the system is incapable of such a leap. As Al Gore, the former Vice President of the United States and 2007 Nobel Peace Prize winner points out,

> "The hard truth is that our economic system is partially blind. It carefully measures and keeps track of the value of those things most important to buyers and sellers [...]. But its intricate calculations often completely ignore the value of other things that are harder to buy and sell: fresh water, clean air, the beauty of the mountains, the rich diversity of life in the forest, just to name a few. In fact the partial blindness of our current economic system is the single most powerful force behind what seem to be irrational decisions about the global environment" (Gore, 2000, pp. 182–183).

---

[1] However, the world economic crisis of 2007–2008 resulted from a breakdown of the market economy that descended into rampant speculation and bidding up of paper assets based on debt from 2001 to 2006. As a result, the U.S. and European Governments interceded in the private sector with trillions of aid dollars to banks and financial institutions, thus implicitly leading to a government-led centralized economy.

[2] Indeed, not only has the market economy contributed to the ecological crisis, but it has created its own fiscal crisis. The market economy requires an abundant money supply, but abundant money leads to speculation and "bubble" formation in paper asset markets leading to the familiar boom, bubble, and bust cycle.

So what is to be done to cure the blindness? How can one ensure that the signs of the biosphere's destruction are properly perceived; not just by a narrow circle of specialists, but by parliamentarians, politicians, business people, civil servants, and others? In this connection it is noteworthy to consider the growing number of voices—not only from pro-Communist or anti-globalization circles—that question the very basis of the market system (i.e., the paradigm of free, unregulated development). Thus, M. Mesarovic and E. Pestel, the authors of the second Club of Rome report entitled *Mankind at the Turning Point*, conclude on the basis of computer models going back 30 years that the natural development of the world's economy under modern conditions is not only irrational, but also downright dangerous (1974).

Undeniably, on the one hand, private enterprise and free competition among producers have already demonstrated their great potential to the world. There is no need to argue that without this mighty engine, modern progress would have been impossible. However, like all natural forces—private enterprise is first and foremost a natural force whose behavior can be predicted or controlled only up to a certain point—this one also contains a creative, as well as a destructive mechanism. The latter can be exemplified at the very least by the periodic overproduction crises that afflicted the Western world prior to the Great Depression of 1929. The cure for this condition appeared only with the introduction of stricter playing rules for business.[3] However, today we are witnessing a crisis that is no longer internal but external with respect to the given system—a crisis that manifests itself ecologically and constitutes a case of overproduction on the part of civilization itself.

As in the fairy tale by the Brothers Grimm entitled *Sweet Porridge* a magic pot has cooked up too much porridge for the planet. The porridge has started to run over the rim of the pot, flooding nearby yards, streets, and in the process—all of our environment. Will modern civilization rein itself in, and limit its appetite without losing its natural qualities? Incidentally, in the Grimm fairy tale, this is accomplished by an external agent—the owner of the magic pot. She has enough time to run into the house and cry out the magic words: "Stop, little pot!" If we continue this analogy, would it not make sense to assume that the market economy also needs some sort of external force that would impose on it a set of external limiting parameters? Without suffocating the system's active and life-giving essence, these external constraints would guide the market into a rationally ordered and safe channel. This is by no means a flight of groundless fancy, as the worldwide calls for new banking regulations demonstrate in the context of the current lending crisis.[4]

---

[3] However, since about 1980 the prevailing view in economics and government is that deregulation spurs growth, and many industries, particularly the banking system, have been deregulated to the point that rampant speculation has become the norm for financial institutions. This led to the savings and loan crisis in the U.S. in the 1980s and 1990s, and the worldwide sub-prime mortgage crisis of 2007–2008. As a result, the stock markets are displaying the same kind of wild rides upward and downward that we saw in 1929–1930.

[4] Actually, the governments of the world, and particularly the U.S., have but one goal in mind, which is to restore the old system by re-inflating the bubble with borrowed money. Re-regulation of the banking system seems unlikely. The almost religious belief in deregulation seems to be too imbedded.

History knows of instances where a liberal economy was quite successful in an environment of strict state control—a form of regulation imposed from the outside. Admittedly, all these cases (except perhaps the New Economic Policy in Communist Russia under Vladimir Lenin in the early 1920s) were associated with martial law. Thus, in the U.K. during World War II the market was centrally regulated. The British economy not only managed to satisfy the country's defense objectives under conditions of a naval blockade, but also ensured a decidedly adequate standard of living for the wartime population. This came into sharp contrast with the living conditions of people in the U.S.S.R. (even in areas very far away from the front) with its centrally planned, collectivized economy.

However, centralization within a single state can be viewed as a mere point of departure for discussing centralization at a different level. This type of centralization would really correspond to the goals of sustainable development. After all, if the latter is viewed as a process of organized retreat by the human race, as Donella Meadows *et al.* interpret it, for example, then one cannot avoid the question of the tools required by such self-organization. Is it possible, for instance, within the U.N. in its current state or does one need another organization endowed with different powers? At any rate, the experience of the past few decades tells us that international financial and economic structures under the control of industrialized countries: for example, the International Monetary Fund (IMF), the International Bank for Reconstruction and Development (IBRD), the World Trade Organization (WTO), as well as the corresponding subdivisions and committees of the U.N. have proved effective only within rather narrow limits and have been ultimately unable to really launch the mechanism of sustainable development.

However, there is one more issue that speaks in favor of a special supranational structure that would take on the burden of responsibility for the state of the Earth's biosphere. In economics there is a well-known concept known as natural monopolies. An example would be the railways, water supply systems, the postal service, or electric networks. By their very nature, such spheres of the economy need to be managed by one agency according to a unified technical policy. The negative results of privatization drives in many parts of the world (e.g., the United Kingdom under Margaret Thatcher and thereafter) speak for themselves.[5] Why then would one not apply the same criteria to the environment that is essentially unified in the same way and should therefore not be considered an "internal matter" (to draw an analogy with the rhetoric of some states regarding human rights) falling under the jurisdiction of an administrative territorial unit or a state structure as a whole?

Current reality is, however, very different. The overwhelming majority of countries have traditionally viewed the environment within their borders not as an integral part of the biosphere as a whole, but rather as their sovereign holdings. This is reminiscent of severely outdated schoolbook political geography. It only remains to guess whose and how many interests are served through the implementation of such

---

[5] It is noteworthy that in the U.S., the drive toward privatizing these natural monopolies has continued unabated since 1980, almost always with disastrous results. The "Enron" experience was perhaps the worst example.

"sovereign' eco-policies. Of course this approach has little to do with what can be termed as true sustainable development. And so, if one stops merely playing at sustainable development and instead tackles the issue in a very serious way, it will be difficult to deny that the world community cannot do without an international and supranational organization equipped with appropriate powers. Such an organization would act on behalf of many different States in order to implement a centralized program for the stabilization of the environment. This program would exist first of all for the sake of the biosphere—and by extension for humanity as a whole—rather than for the promotion of individual geopolitical, ethnic, corporate, and other interests.[6]

Perhaps it is premature to describe and specify the functions of such an organization at this point, which is why for now a general outline would be appropriate:

- Ecological monitoring and environmental observation all over the planet.
- The right to veto large-scale technology and projects at odds with environmental interests.[7]
- The development of social and ecological–economic indicators of sustainable development endowed with the force of law.
- The planning of soil reclamation and the re-establishment of natural ecosystems at the state and continent level.
- The establishment of quotas for energy use, greenhouse gas emissions, and the production of other global pollutants for individual territories and states.
- The allotment of moneys for contribution to an international fund for the support of underdeveloped countries in the grips of a demographic, ecological, or food crisis, etc.

The birth of such transnational political entities as the European Union is an indication that in some parts of the world the nationalist obstacles to the creation of the above organization are diminishing. However, this does not affect the preparedness of populations in the industrialized world to take on certain self-limitations without which an acceptable compromise with nature would be impossible. Sacrificing aspects of the living standard achieved so far—even if only its various obvious excesses—would constitute a major psychological impediment. This is especially so if such a step is not dictated by considerations evident to all (e.g., a state of war, an economic crisis, the immediate consequences of a natural disaster, etc.).[8] And so, we face a Catch-22 situation where any national government that

[6] The danger is that since there are many more developing nations than developed nations, the former could conceivably "gang up" on the latter and impose uneven constraints on the former, as in the Kyoto protocols.

[7] Here again, one must wonder about balance between developing and developed nations. How would this supranational organization deal with the Chinese Three Gorges Dam, or the fact that the Chinese build one new coal-fired power plant per week?

[8] With the current world financial crisis in 2008, and financial hardship widely prevalent in the developed world, it seems unlikely that the public mood in these countries would accept further privation as a consequence of furthering ecological progress.

decides under these circumstances to adopt unpopular measures for a quick move to sustainability would inevitably find itself held hostage by the ballot box. This conundrum is the Achilles' heel of democracy that its detractors never tire to point out with glee.

As Jacques Attali (the former head of the European Bank for Reconstruction and Development) argues in this connection, democracies whose leaders cannot afford even temporary unpopularity are likely to face major future problems. One cannot gain a long-range perspective under constant pressure from the poll numbers. Development is paralyzed if decision-makers cannot be future-oriented risk-takers (cited in Sabov, 2007). A striking example of democratic paralysis is what happened recently in northern Germany. In October of 2008, Anja Hajduk, a senator for environmental affairs from the German Green party, gave the "green" light for the construction of a pit-coal power plant by the Vattenfall Company in Hamburg–Moorburg—and this despite the senator's ecological politics, as well as numerous protests against this polluting project. The senator openly admitted that she was going against her own ecological convictions (Veit, 2008) in order to accommodate short-term economic considerations.[9]

In light of such pressures, a democratic government is sure to gain by delegating the vulnerable aspects of its jurisdiction as outlined above to an impartial supra-national agency which would have the capacity to disentangle the tough knots of sustainable development without consideration for corporate or special interests. Clearly such interest groups are probably going to be in constant conflict with the strategic, global goals of sustainable development. Thus, the world community can hardly do without the sturdy reins of the above-mentioned plenipotentiary agency. We are not talking about rigid totalitarianism proper here, but rather a certain firmness and lack of compromise when it comes to preserving the stability of the environment which is directly linked to the survival of humanity. Naturally, the assumption is that each country would voluntarily follow to the same degree the dictates of such a strategy.

References

Gore, A. (2000). *Earth in the Balance: Ecology and the Human Spirit.* Boston: Houghton Mifflin. Available at *http://books.google.com/books?id=QDbNhec98iEC&pg=PP1&dq =Gore,+Al&sig=ACfU3U31uytaA7TfzegvvvGEiXE-WNcuQg*

Meadows, D. H.; D. L. Meadows; J. Randers (1992). *Beyond the Limits: Confronting Global Collapse, Envisioning a Sustainable Future.* Post Mills, VT: Chelsea Green.

Mesarovic, M.; E. Pestel (1974). *Mankind at the Turning Point.* New York: Dutton.

Sabov, D. (2007). "Political gentleman," *Ievreiskaia Gazeta,* Berlin, p. 3.

Veit, S.-M. (2008). *Die Tageszeitung,* October 2. Available at *http://www.taz.de/1/zukunft/ umwelt/artikel/1/gruenes-licht-fuer-steinkohle/*

[9] Dealing with the ecological crisis requires long-term approaches and policies. However, the 2-, 4-, and 6-year election cycles of democracies require a myopic short-term approach to all problems and situations. As a result, democracies do not seem to be able to cope with long-term problems such as the environment, water supply, energy, and food.

19

Sustainable development and the "real human condition"

Changing the world: psychological obstacles—Biocentrism or anthropocentrism?—In search of a global ethos—The natural imperative or a social utopia—"The real human being" facing the ecological crisis—"No one knows the truth ..."

World-changing ideas have succeeded each other since the dawn of history. Early Christianity 2,000 years ago and the Marxist movement of the past two centuries are two prominent examples of attempts to shake everything up in a fundamental way. Similar to each other in the scale of their respective ambitions and the extremely profound influence they had on the fate of humanity, these movements were characterized by their decisive rejection of the *status quo*, which places them partly in the same league as today's radical "green" movements. However, great ideas derive their strength first and foremost from their positive side. In the framework of the above-mentioned analogy, this was the "Kingdom of Heaven" concept, its Christian version being literally in the heavens and the communist variant—on Earth. The communist kingdom collapsed right before our eyes, having failed the test of life and time. The Christian Kingdom of Heaven appears to have been far sturdier although it has lost quite a number of its followers in the past two centuries.[1]

Current ecological activism—which can be viewed as an extension of the Enlightenment—is also a world-changing movement, but one not based on internal *emotional* convictions or dogmas, as was almost always the case with the

---

[1] It is noteworthy that Christianity had an easier challenge than communism. Christianity did not need to produce anything while communism had to compete with market economies. Both ideologies used force to impose their philosophies on the populace.

transformational teachings of the past. Instead, it rests on the solid foundation of natural science stemming from empirical research across the globe. This research tells us that the haphazard development of the past two to three centuries is bound to precipitate humanity into a global disaster. And yet past experience tells us the opposite, which is perhaps the biggest obstacle on the road to sustainable development.

For what it's worth, the human race has been managing to cope with all its woes and, most importantly, we have constantly pushed forward scientific and technological progress. The latter has brought benefits that our ancestors could not have imagined. And so, when it comes to alarming ecological predictions, one is tempted to think back to all the eschatological epidemics and crises of history. Early Christian communities (e.g., St. Paul's circles) expected the world to end any moment, but it didn't. Early medieval Europe waited with baited breath for the Millennium to usher in the end of the world in the year 1000. When that didn't happen, the Pope and the Holy Roman Emperor were very embarrassed. The Bubonic Plague of the 14th century, with its countless victims also looked like the end, but it wasn't.[2] In short there has never been a lack of wars, famines, and cataclysmic natural disasters. And yet here we are—still alive and kicking! Therefore the temptation to view warnings from the ecological community, however "scientific" and well argued they may be, as another case of alarmism is not without precedent. These are the internal or psychological obstacles faced by human civilization, which is why collecting scientific data on the ecological crisis is only part of the equation. Changing minds and hearts is the other.

Of course, like any other well-intentioned movement, ecological advocacy can go to extremes. In the past two or three decades the ecological movement has seen the appearance of numerous schools and streams advocating the notion of *biocentrism*: that is, the idea that all human activity must be made subordinate not just to the interests of environmental protection but the interests of nature itself. "I hope," writes V. E. Boreiko, editor-in-chief of Kiev's *Humanitarian Ecological Journal* "that the modern understanding of wild nature (embodying the values of wild nature) [. . .], the belief in wild nature as a sacred space and something Utterly Different, the urgency of the immediate need to protect wild nature for its own sake—all this is turning into a new, proactive ideology of an environmentalist community capable of saving the remaining areas of wild nature." Boreiko advocates the creation of a radical ecological organization called *Warriors for Wild Nature*—inspired by the American *Earth First* movement—that would be uncompromising in its drive to defend the "rights and freedoms" of wild nature within parks and preserves, as well as beyond them. "It appears that such an organization would follow a religious model to a certain extent. Its activists would profess a kind of 'environmentalist faith,' viewing wild nature as a sacred domain" (Boreiko, 2002). This resembles very much the initiation ceremony of the *Earth First* movement that in turn looks

---

[2] We may also remember the hysteria surrounding the advent of year 2000 and predictions of disaster for accounting systems.

suspiciously like a religious cult. The *Earth First* ceremony requires that the would-be young initiate dismantle a bulldozer or some other "noxious" piece of technology down to the last screw. Indeed, one can love nature; one can be in awe of it, but deifying nature would take the rational people of the 21st century back to the mythological mindset of our pagan ancestors.

This "stone-age" understanding of biocentrism would require humanity to do the impossible. However anthropocentrist this may sound, the transformation of the environment must take place in relation to human needs, labor, and goal-oriented practical endeavors. Like it or not, anthropocentrism has always been and is likely to remain a dominant human characteristic: that is, we will always pursue our own particularly human interests. On the other hand, primitive anthropocentrism, which declares the human being to be "the pinnacle of creation", is indeed no longer acceptable and must undergo fundamental reassessment. In order to adopt a rational position and not undermine our own existence, we have to start viewing our place on Earth in much more modest terms. We can let philosophers decide what to call this idea: bioanthropocentrism or something else? The point is that such a shift in thinking cannot take place within an artificially created faith, but rather through the transformation of an outdated ethical system that no longer corresponds to today's reality. This too would require great effort and enormous spiritual exertion.

In case there are some who doubt humanity's ability to change its basic values, let us recall that we have passed through the furnace of ethical/ideological transformation more than once in our history: for example, the move from human sacrifice to the rejection of this practice, from paganism to monotheism, from a heliocentric universe to a geocentric one, from Medieval Scholasticism to Renaissance Humanism and then on to Enlightenment, etc. The capacity for such internal restructuring often determined the viability and existence of entire tribes, cultures, and civilizations. This cultural and moral selection process has shaped the face of the modern world in many ways. The survivors have been those nations and peoples that possessed the ability to change their very psychology and rise creatively to evolving natural or man-made challenges. In the same way, today's generation is facing an unprecedented ecological challenge, and the need to reshape our ethics in relation to environmental conservation is dictated to us externally (i.e., on the basis of an objective reality rather than something thought up in an office). As Gumiliov (2001) notes in this connection, humans adapt to natural conditions differently from the way other species do it: namely, when populating a new area, we change not our anatomy or physiology, but rather our mode of behavior.

The need for collective moral renewal is a notion going as far back as biblical times. Thus, in the Old Testament we read many calls for renewal, starting with the Ten Commandments and going through a whole series of prophets. For example "Like a cage full of birds, their houses are full of treachery; therefore they have become great and rich, they have grown fat and sleek. They know no limits in deeds of wickedness; they do not judge with justice the cause of the orphan, to make it prosper, and they do not defend the rights of the needy" (Jeremiah 5:27–28). Accusations of this type often stemmed from a vaguely sensed rift between changing conditions of life and the current ethical system. Such an outdated system's prohibitions or precepts were becoming objectively unattainable or even harmful. In this case, harm and benefit would be interpreted first and foremost as ethical categories (i.e., individual manifestations of good and evil). In the end the application of an old ethical system to a new reality exposed the system's falseness (again, perceived mainly in a moral framework), which served as a stimulus for the formation and crystallization of a new ethical code. This is the essential outline of systemic ethical transformation in the process of progressive human development, and the current crisis fits into the same basic pattern. However, in some ways today's situation is fundamentally different from all that preceded it.

The first difference is the global nature of the new ethical system in question. Various religions have had their own moral systems and values rooted in the traditions of specific human communities. These ideological boundaries are no longer valid. Humanity can survive only if most of the world's peoples can accept the ideals and values of sustainable development. Another distinguishing feature is the way in which the new ethical system combines a moral and a scientific dimension. Perhaps for the first time in history, a common ground for two seemingly disparate notions has been found. A nature-based imperative is meant to make us reassess our ethical vision so that a new, ecological form can be given to the idea of starting life anew.

In 1946, a Russian émigré periodical called *Novyi Zhurnal* published an article by S. L. Frank—a famous Russian Christian philosopher. It was entitled "The heresy of utopianism" and dealt with the temptation of a utopian world order which appeared to have triumphed definitively on the territory of the U.S.S.R. and to have implanted solid roots in Soviet satellite states. The author was hardly counting on being heard by the Soviet leadership. Instead, his text was addressed to certain Western intellectuals under the spell of communist hypnosis that had just been exacerbated by the recent Soviet victory in World War II.

> "The point is that the very mechanism of human existence (its social dimension) is in many of its general manifestations . . . the expression of the way in which human beings are subordinate to the cosmic order . . . Any attempt to change or eliminate these general forms of human life . . . constitutes the expression of an unjust and unnatural human pride—man's titanic drive to build an entirely new world by human efforts alone . . . Social reform is required and must be the subject of rational inquiry because such change creates better conditions for man's inner spiritual re-education. However, in order to achieve this, social reform must take into account the real human condition rather than strive to change it by force" (Frank, 1946).

The real human condition . . . It must be admitted that this phrase was the constant inspiration for the conception of the present chapter. This is the unknown element in a chain of arguments that will, in the final analysis, determine the success or failure of sustainable development. On this basis the large-scale ideas in question will be interpreted either as utopian or realistic. In fact who is to say which aspects of "real humanity"—the stagnant or the flexible, the rational or the blindly passionate and instinctive, the petty egoistical or the selfless—will be manifested in the face of the ecological challenge? At any rate, as the authors of *Beyond the Limits* point out, no one knows the answer. There are reasons these technology-oriented and computer-oriented writers, accustomed to bolstering their conclusions with mathematically tested models, at the end of their books put forth a whole series of questions in a humanitarian manner so atypical for their profession: "Is any change we have advocated in this book, from more resource efficiency to more human compassion, really possible? Can the world actually ease down below the limits and avoid collapse? Is there enough time? Is there enough money, technology, freedom, vision, community, responsibility, foresight, discipline, and love, on a global scale?" (Meadows *et al.*, 1992, p. 234). In all likelihood these questions, as the authors themselves caution, are for now without answers.

References

Boreiko, V. E. (2002). "Wild nature can be saved only by ecological ethics," *Gumanitarnyi ekologicheskii zhurnal* **4**, 48–49 [in Russian].

Frank, S. L. (1946). "The heresy of utopianism," *Novy Zhurnal* **XIV** [in Russian].

Gumiliov, L. N. (2001) *Ethanogenesis and the Earth's Biosphere*. Moscow: Rol'f [in Russian].

Meadows, D. H.; D. L. Meadows; J. Randers (1992). *Beyond the Limits: Confronting Global Collapse, Envisioning a Sustainable Future*. Post Mills, VT: Chelsea Green.

Peccei, A. (1977). *The Human Quality*. New York : Pergamon Press.

20

The social premises of sustainable development and the globalization problem

Authoritarian temptation—Human rights vs. responsibilities—Globalized fragmentation—Common ground: biospheric globalism

Given the historical development of Soviet society, the background of this book's authors may lead one to conclude that there is a utopian subtext behind these pages— a call to create a society of idealists who will happily sacrifice everything for sustainable development. Indeed, the first two decades after the Bolshevik Revolution of 1917 were animated by lofty ideals and the quest for a better place, as well as a better human being. This was a time when poetry, such as the following, expressed the dreams of a whole generation that was willing to give up a great deal for the great communist cause:

> "Already secret trains again
> are chugging to our gloomy borders,
> and Communism seems as close
> as when we heard great Lenin's orders."

This verse was written by the 22-year-old Mikhail Kul'chitsky as the U.S.S.R. was about to face the murderous showdown with Nazi Germany in 1941, and the borders of the communist haven from the "evil world" had to be protected at all cost. Needless to say, the noble aspirations of the first post-revolutionary generation were paid for with the blood and suffering of millions—first under Lenin and even more so under Stalin. It turned out that the means for the creation of the Soviet utopian project annihilated its goals (i.e., first the very foundations of society had to be destroyed and then ...). Then it all began to fall apart until everything vanished in a puff of cynicism and disappointment in 1991 when Mikhail Gorbachev announced the dissolution of the Soviet Union.

Ever since the time of Plato, there has been a temptation to assume that authoritarian and totalitarian regimes "get things done" while democracies chatter away and do nothing. As we read in Plato's *The Republic*, the wisdom of the philosopher king entitles him to arrogate all power since his decisions are based on the eternal forms or truths that make him infallible:

> "Unless philosophers become kings in our cities, or unless those who now are kings and rulers become true philosophers, so that political power and philosophic intelligence converge, and unless those lesser natures who run after one without the other are excluded from governing, I believe there can be no end to troubles, my dear Glaucon, in our cities or for all mankind. Only then will our theory of the state spring to life and see the light of day, at least to the degree possible" (Plato, 1985 [473], p. 165).

And so it may appear that a totalitarian regime, with its concentration of power in one enlightened mind, is in the best position to make sustainable development a reality—in line with Plato's world of perfect forms. However, in the Soviet case the philosopher king turned out to be Stalin who tormented and butchered "those lesser natures" (to borrow Plato's term) for the sake of power—the real purpose of the totalitarian system. As for nature and ecology, we have already demonstrated the Soviet system's view of the natural world as something merely to be conquered and tamed. And so, as history has taught us, authoritarian regimes pursue only one kind of sustainability—their own. The ecological problem cannot be resolved through the sacrifice of freedom and individuality.

Therefore, there is every reason to believe that democratic civil society—characterized by an ongoing dialogue between power and the population, the admission that any given state of things (and any given state) can and should change all the time, the opportunity given to the individual to make conscious choices regarding political power and, most importantly, the notion of individual rights—is the only sociopolitical institution capable of finding the way out of the dead end in which the planet finds itself. However, it is precisely the question of rights that has created a daunting paradox with respect to humanity's place in the biosphere. In 1789, the year of the French Revolution, the National Assembly of France produced *The Declaration of the Rights of Man and Citizen* (n.d.) that spoke of man's natural rights, including the right to liberty and property:

> "Liberty consists in being able to do anything that does not harm others: thus, the exercise of the natural rights of every man has no bounds other than those that ensure to the other members of society the enjoyment of these same rights [...] Since the right to Property is inviolable and sacred, no one may be deprived thereof" (*Declaration*).

After several centuries of gathering momentum, the middle class finally claimed its "rightful" place, effectively inventing the very idea of human rights. The problem is that these rights have come at a cost that is beginning to dawn on us only now. Left out of the liberating equation was the environment: the resource upon which the individual's rights are based. The sacrosanct right to property that has been driving the boundless economic growth of the Western democracies turns out to be limited. And the idea that "the exercise of the natural rights of every man has no bounds" (see above) is just as utopian as the Soviet communist dream.

This book has been considering the question of reassessing the values of modern civilization. We have proposed to reassess our understanding of what constitutes the quality of life, the notion of political sovereignty, the idea of growth, and a number of other fundamental aspects of the human condition. The reassessment of the meaning behind human rights is part of this series. Without questioning the basic principles underlying the concept of human rights, we propose to reconsider its modalities. In this connection it is worthwhile to consider a document adopted in the year 2000 by a commission of UNESCO, entitled *The Earth Charter*. Its elaboration involved hundreds of authoritative specialists, as well as prominent political and social leaders, representing intergovernmental and non-governmental organizations, different social strata, cultures, and beliefs from all the continents. The document offers no magic recipe for resolving the current crisis, and no step-by-step guidelines toward sustainable development are given. However, the value of *The Earth Charter* lies in the *principles* it offers for the solution of our problem.

After pointing out the symptoms of the global illness—planet-wide ecological devastation, resource depletion, growing social strain, increasing violent conflict, the intensification of poverty, and mounting demographic pressure on the environment—*The Earth Charter* calls on humanity to "join together [and] bring forth a sustainable global society founded on respect for nature, universal human rights, economic justice, and a culture of peace [. . .] Fundamental changes are needed in our values, institutions, and ways of living." The notion of human rights is still included, but of particular significance is the reassessment of our values and especially the following point made about collective responsibility outside the framework of the nation-state: "We are at once citizens of different nations and of one world in which the local and global are linked. Everyone shares responsibility for the present and future well-being of the human family and the larger living world" (*The Earth Charter*: Preamble).

The key word here is *responsibility*, and in *The Earth Charter* this idea goes far beyond what the 18th-century ideologues of the French *Declaration of the Rights of Man and Citizen* had in mind. Whereas the latter were affirming human rights in the face of human enslavement, *The Earth Charter* affirms human responsibility in the face of the arrogance produced by the notion of human rights. This responsibility is that of the individual and whole peoples before the biospheric community—the heritage to be left for future generations. After 200 years of movement toward liberation, we are discussing not a counter-movement but rather a lateral shift—a qualitative revision in thinking. The implicit idea of *The Earth Charter* is that the legal and political changes required to implement its principles can take place only after the corresponding mental changes have taken place all over the world. And despite the gloomy picture we have painted, these mental changes appear to be moving forward.

However, the post–Cold War period brought new obstacles to the emergence of a global responsible consciousness. As strange as it may seem, the confrontation between the Eastern Bloc and the West split the world and yet, at the same time imposed a kind of global responsibility. The sense of belonging to one camp or the other, espousing "our" ideology and rejecting "their" dogma, brought people together across borders, cultures. and oceans. To quote Panarin (1994): "The paradox of modern history is that the bipolar structure of the world—despite its horrific excesses stemming from the mindlessly wasteful competition of two super-powers—was still more in harmony with the expectations of a civilized mind-set with its understanding of world order, stability, predictability, and 'unified spaces.' Today's lack of structure is threatening to wipe out all sense of order" (Panarin, 1994).

The modern identity crisis has been analyzed by the American political scientist Samuel P. Huntington. In his widely quoted article entitled "The clash of civiliza-tions," Huntington writes about the fragmentation of the world in the post-Cold War period and points out:

> "... as people define their identity in ethnic and religious terms, they are likely to see an 'us' versus 'them' relation existing between themselves and people of different ethnicity or religion. The end of ideologically defined states in Eastern Europe and the former Soviet Union permits traditional ethnic identities and animosities to come to the fore. Differences in culture and religion create differ-ences over policy issues, ranging from human rights to immigration to trade and commerce to the environment" (Huntington, 1993).

Deriving their sense of identity from the rejection of Western values, large masses of people (especially the underprivileged) in the developing world cut themselves off from any potential for a sense of global community. Huntington zeroes in on the Moslem civilization as the hotbed of sectarian thinking threatening the stability and unity of the world. To what extent the current wave of Moslem fundamentalist aggression can be identified with the Moslem world as a whole is an issue up for debate, and Huntington has been taken to task for talking of different *civilizations* to begin with (e.g., by Berman, 2003). However, the anarchy, bullets, and bombs marking the landscape of Afghanistan, Pakistan, Iraq, Somalia, Sudan, and other unstable societies—not to mention the successful and foiled attempts to spread the fire to the heartland of the West—constitute a powerful distraction from the concept of global environmental responsibility.

Huntington argues that the imperialist system of the 19th and early 20th century made it possible to impose something like universal (read "Western") values by force in many parts of the world beyond Europe. However, that system no longer exists, and "the belief that non-Western peoples should adopt Western values, institutions, and culture is, if taken seriously, immoral in its implications" (Huntington, 1996a). The old formula for bringing people together is bankrupt, and so far little has appeared to take its place—all the more so because the world is suffering from a lack of leadership. The recent disregard for international norms by the Bush administration[1]—from the invasion of Iraq to Guantanamo—has added to the sense of "sheer chaos" (to use Huntington's term) that has gripped the post-Cold War world:

> "The illusion of harmony at the end of that Cold War was soon dissipated by [...] the resurgence of neo-communist and neo-fascist movements, intensification of religious fundamentalism, the end of the 'diplomacy of smiles' and 'policy of yes' in Russia's relations with the West, the inability of the United Nations and the United States to suppress bloody local conflicts, and the increasing assertiveness of a rising China. In the five years after the Berlin wall came down, the word 'genocide' was heard far more often than in any five years of the Cold War. The one harmonious world paradigm is clearly far too divorced from reality to be a useful guide to the post–Cold War world" (Huntington, 1996b).

Perhaps the only common ground left in this confusing and disjointed world is *biospheric globalism*. This is not "globalization" in the economic or even political sense of the term. This is not even Marshall McLuhan's "global village"—a human community integrated into one whole through a network of increasingly efficient communication. To be sure, these phenomena have indisputably brought people together more than anything else in the past. However, in the dizzying array of values, cultures, and (if we follow Huntington's line of reasoning) civilizations, the only leveler appears to be the realization that the well-being of the biosphere is primary while all our creations and obsessions are only its outgrowths. This is biospheric globalism—an approach to life that can cut the human ego down to size and help us create a list of priorities for the future. We have tried to make a small contribution to this list, joining a choir of voices that have been "calling in the wilderness" to preserve the wilderness. Let us hope that the call does not remain unanswered.

References

Berman, P. (2003). *Terror and Liberalism*. New York: Norton.

Declaration of the Rights of Man and Citizen (n.d.). Paris: Ministry of Foreign Affairs. Available at *http://www.diplomatie.gouv.fr/en/france_159/discovering-france_2005/france-from-to-z _1978/the-symbols-of-the-republic-and-bastille-day_2002/the-declaration-of-the-rights-of-man-and-the-citizen_1505.html*

Huntington, S. (1993). "The clash of civilizations?" *Foreign Affairs* **72**(3), 22–49.

Huntington, S. (1996a). "The West: Unique, not universal," *Foreign Affairs* **75**(6), November/December, 28–46.

Huntington, S. (1996b). *The Clash of Civilizations and the Remaking of World Order*. New York: Simon & Schuster.

Panarin, A. S. (1994). "Challenge," *Znamia* **6**, 149–160.

---

[1] Or the invasion of Georgia by Russia in 2008.

Plato (1985 [473]). *The Republic*. (translated by R. W. Sterling and W. C. Scott). New York: Norton.

UNESCO (2000). *Earth Charter*. New York: United Nations Educational, Scientific, and Cultural Organization. Available at *http://www.earthcharterinaction.org/2000/10/the_earth_charter.html*

Index